Catalysis and Surface Characterisation

Catalysis and Surface Characterisation

Edited by

T. J. Dines
C. H. Rochester
J. Thomson
Department of Chemistry, University of Dundee

The Proceedings of a Meeting of the Royal Society of Chemistry
Surface Reactivity and Catalysis Group held at Dundee University
on 30 March – 1 April 1992.

Special Publication No. 114

ISBN 0-85186-335-3

A catalogue record for this book is available from the British Library

© The Royal Society of Chemistry 1992

All Rights Reserved
No part of this book may be reproduced or transmitted in any form or
by any means – graphic, electronic, including photocopying, recording,
taping or information storage and retrieval systems – without written
permission from The Royal Society of Chemistry

Published by The Royal Society of Chemistry,
Thomas Graham House, Science Park, Cambridge
CB4 4WF

Printed by Redwood Press Ltd., Melksham, Wiltshire

Preface

Catalysis and surface characterisation contains a selection of papers presented at the meeting of the RSC Surface Reactivity and Catalysis Group, held at Dundee University from 30 March to 1 April 1992. The meeting was attended by a hundred and twenty delegates from industrial and academic laboratories throughout the UK, EC and USA.

In the last three decades heterogeneous catalysis has developed from methodology largely relying upon trial and error into a science embracing diverse aspects of chemistry and physics. Central to this progress have been the advances in spectroscopic and other techniques, which have led to a better understanding of surfaces. Industrial applications have provided the major impetus for the improved understanding of catalysts that has resulted from surface characterisation; thus most of the catalysts described in this book are of current industrial importance.

Although FTIR remains the dominant technique for surface characterisation, other spectroscopic methods are being increasingly utilised. The power of NMR as a probe of surface structure and adsorbed species is now being exploited although not many laboratories are as yet equipped with solid state NMR facilities. Other techniques described in this book include X-ray methods, especially EXAFS, electron microscopy and also non-spectroscopic methods such as temperature programmed reduction.

The organisers would like to express their gratitude to the participants in the meeting and to the authors for their cooperation in submitting camera ready manuscripts.

T. J. Dines
C. H. Rochester
J. Thomson

Contents

In situ solid state NMR studies of reactions on
catalysts
(BRITISH VACUUM COUNCIL ANNUAL ADDRESS 1992)   1
J. F. Haw

High thermal stability of nickel-aluminium mixed
oxides obtained from hydrotalcite-type precursors   32
P. Beccat, J. C. Roussel, O. Clause, A. Vaccari
and F. Trifiro

A supported copper catalyst: a combined XAS and XRD
investigation during calcination, reduction, N_2O
decomposition and catalytic reaction   42
G. Sankar, J. W. Couves, D. Waller, J. M. Thomas,
C. R. A. Catlow and G. N. Greaves

Lability of small rhodium particles in carbon
monoxide: influences of chlorine and particle size   51
P. Johnston and R. W. Joyner

The advantageous use of microwave energy in catalyst
characterisation   60
G. Bond, R. B. Moyes and D. A. Whan

Inelastic neutron scattering studies of catalysts –
thiophene adsorbed on sulphided molybdena/alumina   68
P. C. H. Mitchell, J. Grimblot, E. Payen and
J. Tomkinson

The surface chemistry of ceria doped with lanthanide
oxides   76
P. G. Harrison, D. A. Creaser, B. A. Wolfindale,
K. C. Waugh, M. A. Morris and W. C. Mackrodt

FTIR identification and microcalorimetric
characterization of $Mo=CH_2$ and $(\eta\text{-}C_2H_4)Mo^{4+}$
complexes after cyclopropane adsorption on
photoreduced MoO_3/SiO_2 olefin metathesis
catalysts   87
K. A. Vikulov, B. N. Shelimov, V. B. Kazansky,
G. Martra, L. Marchese and S. Coluccia

Infrared spectroscopic studies of chemisorption
on metal surfaces   97
(PLENARY LECTURE)
M. A. Chesters, P. Hollins and D. A. Slater

Thermal evolution and hydrogenation of ethylidyne
on Pt(111) studied by RAIRS   109
G. McDougall and H. Yates

The adsorption and reaction of formic acid on
oxygen-doped silver films: an infrared study   118
S. Munro and R. Raval

In situ **characterization of rhodium hydroformylation catalysts** 127
L. Sordelli, R. Psaro, C. Dossi and A. Fusi

Pretreatment effects on CO oxidation behaviour of platinum-rhodium catalysts - a combined FTIR, XPS and TPR study 136
J. A. Anderson and B. Williams

TEM and *in situ* **FTIR characterization of Pd-Fe particles on alumina: geometric and electronic effects due to iron** 145
E. Merlen, N. Zanier-Szydlowski, P. Sarrazin and M. L. Hely

In situ **solid state NMR study of the adsorption of methanol and water on the $Cu/ZnO/Al_2O_3$ methanol synthesis catalyst** 155
A. Bendada, J. B. C. Cobb, B. T. Heaton and J. A. Iggo

The catalytic decomposition of formic acid by the Cu/Pd[85:15]{110} P(2x1) surface 165
M. A. Newton, S. M. Francis and M. Bowker

Enantioselective hydrogenation: use of low surface coverages of cinchonidine modifier to determine site character in the platinum-catalysed conversion of pyruvate to R-(+)-lactate 174
K. E. Simons, A. Ibbotson and P. B. Wells

An *in situ* **study of pyrochlore-type catalysts for the formation of synthesis gas from methane and CO_2** 184
A. T. Ashcroft, A. K. Cheetham, R. H. Jones, S. Natarajan, J. M. Thomas and D. Waller

Chemically-stimulated exo-electron emission from platinum wires 190
M. E. Cooper

Surface characterization of Pt/TiO_2 catalyst by HRTEM and IR investigation. H_2 and CO adsorption on pre-oxidized sample 196
L. Marchese, G. Martra, S. Coluccia, G. Ghiotti and A. Zecchina

X-ray absorption spectroscopic investigation of microporous crystalline cobalt substituted aluminium phosphates 202
J. Chen, G. Sankar, R. H. Jones, P. A. Wright and J. M. Thomas

Mechanism of ethyne hydrogenation in the region of the kinetic discontinuity: composition of C_4 products 207
R. B. Moyes, D. W. Walker, P. B. Wells, D. A. Whan and E. A. Irvine

A molecular beam, RAIRS and NEXAFS study of the oxidative dehydrogenation of methanol on Cu(110) — 213
R. Davis, S. M. Francis, P. D. A. Pudney, A. W. Robinson and M. Bowker

Ethanol oxidation on rhodium(110) — 221
Y. Li and M. Bowker

Atom-probe field ion microscope study of surface changes in a platinum/rhodium catalyst alloy upon oxidation and reduction treatments — 228
S. Poulston and G. D. W. Smith

Carbon monoxide hydrogenation over alumina supported ruthenium-manganese bimetallic catalysts — 234
T. H. Syed, B. H. Sakakini and J. C. Vickerman

SIMS and EELS study of the adsorption of CO on Mn/Ru(0001) — 242
T. H. Syed, B. H. Sakakini and J. C. Vickerman

An electron microscopy study of the sintering of Pt/Rh/alumina/ceria catalysts — 249
A. Cook, A. G. Fitzgerald and J. A. Cairns

EXAFS characterization of surface species in Na-doped molybdena-titania systems — 255
P. Malet, A. Munoz-Paez, C. Martin, I. Martin and V. Rives

X-ray photoelectron spectroscopic determination of the surface oxidation state of ceria in UHV — 261
P. G. Harrison, D. A. Creaser, B. A. Wolfindale and M. A. Morris

Characterisation and testing of carbon monoxide oxidation catalysts — 269
N. Jorgensen

Oxidation studies on Pt-Sn supported catalysts — 276
J. W. Curley and O. E. Finlayson

Characterization by IR and Raman spectroscopy of Pt/alumina catalysts containing ceria — 282
M. S. Brogan, J. A. Cairns and T. J. Dines

Subject Index — 289

In Situ Solid State NMR Studies of Reactions on Catalysts

James F. Haw
DEPARTMENT OF CHEMISTRY, TEXAS A&M UNIVERSITY,
COLLEGE STATION, TX 77843, USA

1 INTRODUCTION

A major emphasis of my research group for the past several years has been the development of solid state NMR methods for the study of chemical reactions on heterogeneous catalysts. Most of this work has been directed toward in situ methodologies in which the reactions are studied while in progress in the NMR probe at some appropriate temperature. Using such methods, we have studied catalytic reactions such as olefin oligomerization,[1-3] the chemistry of acetylene,[4,5] hydrocarbon cracking,[6,7] the methanol-to-gasoline process which is catalyzed by zeolite HZSM-5,[8,9] and methanol synthesis on $Cu/ZnO/Al_2O_3$.[10] All of these studies were carried out using the line-narrowing technique magic-angle spinning (MAS) in order to obtain high resolution ^{13}C NMR spectra of the reacting systems. Recently we have also applied one and two dimensional proton ^1H MAS NMR in an in situ mode to study the interactions of adsorbates with strong acid sites in zeolites.[11,12]

In situ MAS NMR studies of catalytic reactions face several conflicting sample handling requirements. Most catalyst-adsorbate samples are air or moisture sensitive, and such samples generally require preparation on a vacuum line. The integrity of the sample must be maintained throughout all subsequent steps, including loading into the MAS rotor. This can be especially challenging if the reactant is only weakly adsorbed. Furthermore, some of the catalyst-adsorbate systems of interest are reactive at ambient or subambient temperature, and sample preparation and handling must then be carried out at cryogenic temperature to present the NMR experiment with an unreacted sample.[1-3] Regardless of the preparation and handling requirements, the sample must end up inside an MAS rotor with a sample compartment that is ca. 1 cm in length by 4 to 6 mm in diameter. During spectral acquisition, the rotor spins at a rate of typically 3,000 to 10,000 rotations per second while suspended on air bearings. In order to spin properly, the sample must be well balanced.

There are at least two general approaches for reconciling the conflicting requirements of maintaining sample integrity while loading it into a MAS rotor. The first approach makes use of glass ampules prepared by flame sealing on a vacuum line.[13] The second approach for preparing samples for in situ MAS NMR studies makes use of rotor caps with 10 to 15 deformable ridges that create an air-tight seal when the cap is driven into the rotor.[1,14] This sealing methods works well for temperatures up to 523 K, but glass ampules are generally required for higher temperatures, especially if high pressures are generated. Using rotor designs based on grooved caps, the CAVERN apparatus was developed for preparing highly reactive catalyst-adsorbate samples for in situ MAS NMR studies.[1] The somewhat contrived acronym originally stood for "cryogenic adsorption vessel enabling rotor nestling." We now use the term CAVERN to designate any device for sealing or unsealing an MAS rotor on a vacuum line. Indeed, the original CAVERN apparatus proved useful for general preparation of samples at room temperature for in situ studies at elevated temperatures. Using the CAVERN, adsorption onto the catalyst takes place inside the MAS rotor, which is then sealed prior to removal from the vacuum line and transfer to the NMR probe. Figure 1 shows a drawing of the second-generation CAVERN apparatus.[15]

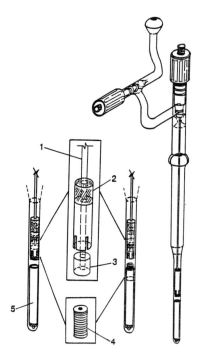

Figure 1 Second-generation CAVERN apparatus (right) for both capping rotors following adsorption of reactants and the capping and uncapping of rotors to allow the sequential adsorption of reactants onto catalysts. The Kel-F plunger mechanism used for the capping of rotors is shown in this diagram. The inset is a view of the lower section of the CAVERN apparatus (left) that depicts the capping procedure. Left and right diagrams show the lower section of the CAVERN prior to capping and immediately after capping, respectively; and the middle sections show enlarged views of the plunger mechanism and Kel-F cap used. See reference 15 for part numbering.

2 APPLICATIONS OF IN SITU NMR

In situ NMR studies of catalytic reaction mechanisms will be illustrated with experiments excerpted from studies of three processes: catalytic cracking, methanol to gasoline chemistry, and methanol synthesis on $Cu/ZnO/Al_2O_3$. All of the studies reported here were carried out using ^{13}C MAS NMR at field strengths of either 4.7 or 7.05 Tesla.

In Situ MAS NMR Study of Cracking Reactions

The catalytic cracking of crude oil into gasoline-range hydrocarbons on zeolite catalysts is responsible for a large portion of world petroleum production. Figure 2 shows the ^{13}C MAS NMR spectra of the reactions of ethylene oligomers on zeolite HZSM-5.[6] The room-temperature spectrum consists of a broad peak centered at 32 ppm assigned to $(-CH_2-)_n$ chains and a narrower peak at 13 ppm assigned to $-CH_3$ groups. Heating of the sample to 523 K resulted in the cracking of the oligomers to propane, butanes, and higher aliphatic hydrocarbons (10-40 ppm). A small amount of aromatics was observed (130-140 ppm), as well as a peak at 250 ppm due to cyclopentenyl cations (vide infra). A general shift towards lower molecular weight hydrocarbons (propane and butanes) and more aromatics occurred when the sample temperature was increased to 573 K. At 623 K the products consisted almost entirely of methane (-10 ppm), ethane (5 ppm), propane (15 ppm), and methyl-substituted benzenes (125-140 ppm).

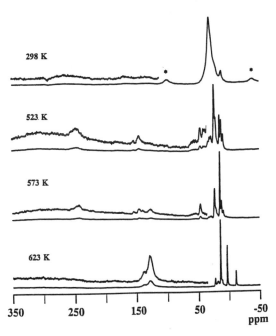

Figure 2 ^{13}C MAS NMR spectra showing the cracking of ethylene-$^{13}C_2$ oligomers on zeolite catalyst HZSM-5. As the sample temperature was increased, the hydrocarbon product distribution shifted from medium range hydrocarbons (C_3-C_6) to light hydrocarbons (C_1-C_3) and coke. Highlighted regions show the presence of methyl-substituted cyclopentenyl carbenium ions (250, 148, and 48 ppm) in the spectra at 523 and 573 K. * denotes spinning sideband.

These results are consistent with the expected cracking chemistry of ethylene oligomers in zeolite catalysts.[16] At 523 K cracking reactions produced aliphatic hydrocarbons and carbenium ions. No ethane or methane were generated at 523 K, suggesting that ß-elimination to form C_3 and higher hydrocarbons was the predominant cracking reaction. When the sample was heated to 573 K, the average carbon chain length decreased as secondary cracking reactions began. Methane and ethane did not appear until the sample was heated to 623 K. The observation of these species demonstrates that different cracking chemistry is occurring on the zeolite at 623 K than at 523 K. For example, in the previous study of the cracking reactions of propene oligomers on HY at 503 K,[6] we observed no methane. In this study the formation of methane can be explained by mechanisms that involve the formation of penta-coordinated carbonium ions which can eliminate methane.

The peak at 250 ppm is indicative of three-coordinate carbenium ions.[17] We have previously observed a resonance at 250 ppm in our study of the oligomerization of propene on zeolite HY at low temperatures and assigned it to alkyl-substituted cyclopentenyl cations.[1] In that paper, the 250-ppm resonance was assigned to the C_1 and C_3 carbons in 1,2,3-trimethylcyclopentenyl cation (I), with a peak at 158 ppm due to the C_2 carbon. In this study we do not observe a resonance at 158 ppm, but rather a resonance at 148 ppm that correlated with the appearance and disappearance of the peak at 250 ppm. The ratio of the integrated intensity of the peak at 250 ppm to the peak at 148 ppm is approximately 2:1. This result suggests that the resonances are due to 1,3-dimethylcyclopentenyl cation (II). The solution-state ^{13}C chemical shifts of 249 ppm for the C_1 and C_3 carbons and 148 ppm for the C_2 carbon for II agree with this assignment.[18] Further evidence for the existence of this cation is the resonance at 48 ppm. This resonance is outside the normal chemical shift region for aliphatic hydrocarbons, but is in very good agreement with the reported value of 48.7 ppm for the C_4 and C_5 carbons of II. It is known that cyclopentenyl cations are very stable carbenium ions.[19] Cyclopentenyl cations have long been proposed to form in the reactions of ethylene on zeolites. From studies of the reactions of ethylene on rare-earth exchanged X-type zeolites performed in the late 1960's, Venuto proposed a mechanism that involved cyclopentenyl cations as intermediates in the production of aromatic hydrocarbons.[16] The above observation is consistent with that proposal.

The Methanol-To-Gasoline Process: Selected Results From a Work in Progress

One of the most successful routes to synthetic fuels

is the conversion of methanol to gasoline (MTG) using the zeolite catalyst HZSM-5. This process, developed by Mobil, now produces one-third of New Zealand's gasoline supply. Although this process has long been studied, the reaction mechanism is still poorly understood. Key questions in MTG chemistry include the mechanism of initial C-C bond formation, the identity of the "first" olefin, and the reason for an observed induction period preceding hydrocarbon synthesis.[20,21]

Some of the proposed mechanisms for C-C bond formation can be organized into the following classifications: carbene,[22,23] carbocation,[24] oxonium ylide,[25,26] and free radical.[27] Some of these mechanisms[24-26] invoke surface methoxyl species as methylating agents in the reaction. Dybowski and coworkers have recently reported a ^{13}C NMR study which supported the formation of methoxyl groups in HZSM-5,[28] but the existence of these species is still controversial.[29,30] Most of the above mechanisms predict intermediary roles for ethylene and/or ethyl methyl ether. Mixed evidence for oxonium routes has come from our recent NMR spectroscopic observation of the formation of trimethyloxonium from dimethyl ether on HZSM-5.[9] Two other mechanisms that have been proposed recently involve CO as either an intermediate (hydrocarbonylation)[31,32] or catalyst (ketene).[30]

We have carried out a number of in situ MAS NMR experiments in an attempt to understand the MTG reaction mechanism. The strategy has been to study the chemistry of methanol, dimethyl ether, and a number of proposed intermediates and suitable analogs, alone or in combination. The reaction mechanism remains unknown as this chapter is written. The examples selected for inclusion here give an overview of this work and illustrate in situ experiments involving multiple adsorbates or reactants. This work also demonstrates that it is sometimes possible to distinguish between reagents adsorbed into a zeolite and the same species in the gas phase.

All samples were prepared and reacted using the CAVERN apparatus unless stated otherwise. In several of the experiments presented here, the sample was heated at a rate of ca. 1 K/s in the MAS NMR probe to a final temperature of 523 K, and then ^{13}C spectra were acquired over the time course of the reaction. This protocol facilitates comparisons of reaction kinetics. In other experiments reported here, the temperature was increased in a step-wise manner in order to show more clearly a sequence of reaction steps. All chemical shift assignments were made by comparison to literature values. For some of the observed intermediate species, these assignments were confirmed by the observation of multiplet patterns in the absence of proton decoupling and/or the direct adsorption and study of these species in the zeolite. For convenience, all of the relevant chemical shift assignments are compiled in Table 1, and there will

Table 1 ^{13}C chemical shift assignments[a] for reactants, products, and the intermediates observed in the in situ studies of MTG chemistry reported here.

Species	Observed ^{13}C chemical shift[b]
Methane	-10
Ethane	5
Methanethiol	8
Isopentane	10, 21, 30, 31
2,2-dimethylbutane	10, 28, 29, 36
3-methylpentane	10, 18, 29, 36
2-methylpentane	13, 20, 22, 28, 41
n-butane	13, 25
n-pentane	14, 23, 35
Propane	14, 15
Ethyl methyl ether	14, 57, 71
Diethyl ether	14, 67
1,2,3,5-tetramethylbenzene	15, 21, 131, 133
Ethanol	17, 63
Dimethyl sulfide	18
1,2,4,5-tetramethylbenzene	21, 131, 133
Isobutane	23, 24
Trimethylsulfonium	28
Methanol (exogenous)	48
Methanol (adsorbed)	50
Dimethyl ether (exogenous)	58
Dimethyl ether (adsorbed)	60
Trimethyloxonium	80
Ethylene	122
CO_2	126
Formic Acid	165
CO	184

[a]Based on literature and on multiplicity patterns obtained without proton decoupling.
[b]ppm from TMS

be little mention of chemical shifts in the text.

Roles of Water   Figure 3 shows selected ^{13}C MAS NMR spectra from an in situ study of the reactions of methanol-^{13}C on a sample of HZSM-5 that had been activated to a final temperature of 623 K. An equilibrium between methanol, dimethyl ether and water was set up immediately after heating to 523 K in the NMR probe. No further reaction occurred during an induction period of ca. 30 min, followed by the start of hydrocarbon synthesis. Ethylene and methyl ethyl ether were observed at the onset of hydrocarbon formation, and continued to be in observable concentrations until the reaction neared completion after ca. 70 min at 523 K. No methanol, dimethyl ether, ethylene or ethyl methyl ether remained at that time. No changes in the composition were observed after the sample was returned to room temperature (spectra not shown), i.e., the reactions were irreversible. The major products of the reaction were isobutane and propane, with minor products consisting of butanes and pentanes as well as aromatics. This product distribution reflects secondary cracking reactions.

Figure 4 shows results from a study identical to the above one except that the catalyst was activated to a higher temperature, 673 K vs. 623 K. The experiment depicted in Figure 4 differed in several significant ways. The induction period was ca. twice as long for the catalyst activated at the higher temperature, and the rate following the induction period was sli-

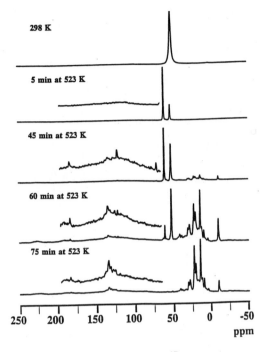

Figure 3 In situ ^{13}C MAS NMR spectra showing the reactions of methanol-^{13}C on zeolite HZSM-5 that was activated to 623 K. The downfield region is expanded to show the formation of ethylene, ethyl methyl ether, and CO. Resonances for all species in this and other MTG studies are assigned in Table 1.

ghtly slower. Furthermore, ethylene and ethyl methyl ether did not form in observable steady-state concentrations.

The effects of catalyst activation at 673 K could be reversed by the coadsorption of water. An experiment identical to the previous one except that 7.5 mmol of water per gram of catalyst was coadsorbed was also performed (not shown). In this case, the induction period at 523 K was shortened to ca. 40 min and small amounts of ethylene and ethyl methyl ether were observed while hydrocarbon synthesis was in progress. Thus the effects of activating at higher temperature could essentially be reversed by coadsorption of water.

Thermogravimetric and infrared studies have previously shown that water is still evolved from HZSM-5 at tem-

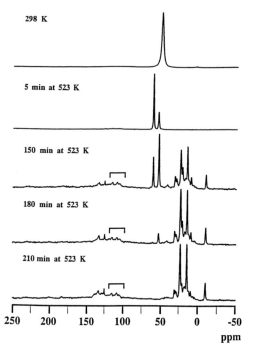

Figure 4 In situ ^{13}C MAS NMR spectra showing the reactions of methanol-^{13}C on zeolite catalyst HZSM-5 that had been activated to 673 K. The spectra were acquired after the sample had been maintained at 523 K for the indicated time. Brackets denote background signals from Kel-F caps.

peratures greater than 773 K.[33] The experiments depicted in Figure 4 show that the residual water has a significant effect on the kinetics of hydrocarbon synthesis at 523 K. Water is a coproduct in the MTG reaction. These studies suggest that the water content in the zeolite might in part explain the observed induction period. Lower activation temperatures and/or the coadsorption of water also lead to increases in the steady-state concentrations of ethylene and ethyl methyl ether.

Water significantly modifies the acidity of the zeolite. Leveling effects are well known in acid-base chemistry in aqueous and nonaqueous solutions. Figure 5 shows that leveling effects are also important in acidic zeolites. The cracking of ethylene oligomers on zeolite HZSM-5 was discussed in detail in the previous section of this account. The reactions of ethylene are very different in the presence of water. 7.5 mmol/g of water was uniformly diffused throughout the catalyst, and then 1.0 mmol/g of ethylene-$^{13}C_2$ was adsorbed. As suggested in Figure 5, the ethylene was unreactive at room temperature. Upon heating to 433 K, most of the ethylene hydrolyzed to ethanol, and this mixture readily synthesized aliphatic hydrocarbons at 523 K.

Adsorbed Reactants Versus Gas Phase Reactions. It is important to differentiate between species adsorbed in the zeolite and gas phase (exogenous) species. In all of the experiments reported here, the species were primarily adsorbed in the zeolite channels, but it is possible to create conditions in which exogenous species are also observed. Figure 6 shows an example of this. Methanol-^{13}C was adsorbed on HZSM-5, and the sample was sealed in a

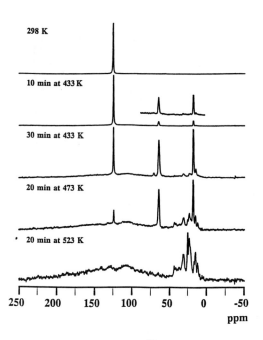

Figure 5 In situ ^{13}C MAS NMR spectra showing the reactions of ethylene-^{13}C$_2$ and water on HZSM-5. Water was adsorbed first to prevent oligomerization of ethylene. The highlighted region of the spectrum taken at 433 K shows the formation of ethanol.

glass ampule such that the zeolite filled only ca. ½ of the volume. At room temperature the methanol was completely adsorbed in the pores of the catalyst, as is shown by the broad resonance in the room-temperature spectrum. When the sample was heated to 523 K, approximately 50% of the methanol and dimethyl ether diffused out of the zeolite, resulting in two peaks for both methanol and dimethyl ether. A third spectrum was acquired at 623 K immediately before hydrocarbon synthesis. This spectrum also clearly reveals adsorbed vs. gas phase species. We attribute the observation of exogenous and adsorbed methanol and dimethyl ether in this experiment to the large gas volume above the catalyst. The 5-ppm difference in chemical shift for methanol and 3-ppm difference for dimethyl ether for the adsorbed verses gas phase species at 523 K is rationalized as due to hydrogen bonding in the adsorbed phase. In future experiments, the ability to differentiate between exogenous and adsorbed species may be useful for studying diffusivity of adsorbates in and out of the zeolite at various temperatures.

Possible Roles of CO. As mentioned previously, reaction mechanisms have been proposed for MTG chemistry that involve CO as either an intermediate or a catalyst. These are reasonable proposals as CO often forms in low concentrations during MTG reactions. These proposals were investigated by performing in situ experiments with added CO, which was generated in situ in the catalyst by thermal decomposition of formic acid. Parallel experiments with or without CO exhibited similar rates. The experiments depicted in Figure 7 address the proposed intermediary role. The top four ^{13}C MAS spectra in Figure 7 show the results of heating methanol-^{13}C and formic acid-^{13}C. CO formation from formic acid was quantitative after 5 min at 523 K. The methanol and dimethyl ether reacted to form hydrocarbons while the CO was partially converted to CO_2 by the water-gas shift reaction. The total integrated intensity of CO and CO_2 remained constant during hydrocarbon synthesis, suggesting that CO was not incorporated into the products. A more definitive experiment was one identical to the one described above, with the exception that formic acid-^{13}C was adsorbed with unlabelled methanol. The final spectrum at 523 K from that experiment is shown at the bottom of Figure 7. Clearly, no ^{13}C

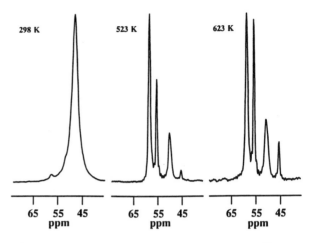

Figure 6 Expanded views from a ^{13}C in situ MAS study of MTG chemistry showing the signals from methanol and dimethyl ether. In the high-temperature spectra two peaks are present for both species; gas phase (48.0 ppm) and adsorbed (53.1 ppm) methanol, and gas phase (58.0 ppm) and adsorbed (61.0 ppm) dimethyl ether.

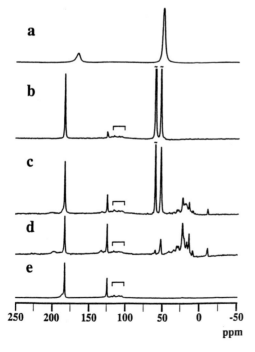

Figure 7 In situ ^{13}C MAS NMR spectra showing the chemistry of formic acid-^{13}C and methanol-^{13}C (a-d) and of formic acid-^{13}C and unlabeled methanol (e) on HZSM-5. (a), At 298 K prior to raising the probe temperature; (b), after 5 min. at 523 K, showing conversion of formic acid to CO; (c), after 150 min. at 523 K; (d), after 210 min. at 523 K. (e), same as (d) except that unlabeled methanol was used, demonstrating that the label from ^{13}CO is not incorporated into the hydrocarbon products. Background signals from Kel-F endcaps which were used to seal the sample rotor are denoted by brackets.

label was incorporated into the hydrocarbon products. A recent mass spectrometry study has arrived at a similar conclusion,[34] and other reaction mechanisms should be studied to explain MTG chemistry.

Oxonium Ion Routes. Several studies have proposed that oxonium ions are key intermediates in the C-C bond forming reaction in either MTG chemistry on HZSM-5 or in analogous reactions on other catalysts (Scheme I).[20,25,26] Trimethyloxonium (III) is deprotonated by an unspecified

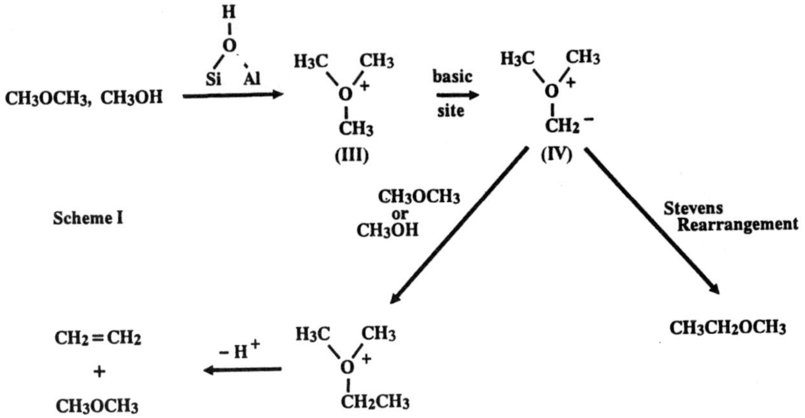

Scheme I

$CH_3OCH_3, CH_3OH \longrightarrow$ (III) $\xrightarrow{\text{basic site}}$ (IV)

(III) → via CH_3OCH_3 or CH_3OH → H_3C-O^+(-CH_3)-CH_2CH_3 → $-H^+$ → $CH_2=CH_2 + CH_3OCH_3$

(IV) → Stevens Rearrangement → $CH_3CH_2OCH_3$

basic site to form an onium ylide (IV). Van den Berg proposed that the onium ylide undergoes a Stevens rearrangement to form ethyl methyl ether,[26] while Olah predicted that the ylide is methylated to form ethyldimethyloxonium and then ethylene and dimethyl ether.[25]

Figure 8 shows an in situ study of dimethyl ether on HZSM-5. The spectrum at 293 K shows that in addition to the dimethyl ether signal at 60 ppm there is a signal at 80 ppm and a smaller shoulder at 50 ppm. The 80-ppm resonance is characteristic of trimethyl oxonium. The balanced reaction for the disproportionation of dimethyl ether on HZSM-5 (Scheme II) requires a stoichiometric number of Bronsted sites and produces one equivalent of

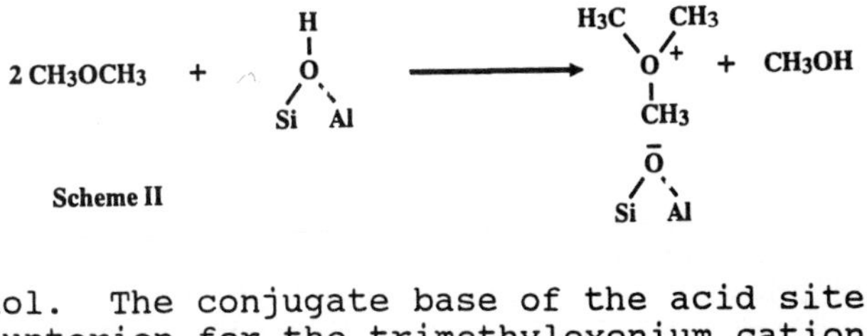

Scheme II

$2 CH_3OCH_3 + \text{H-O-(Si,Al)} \longrightarrow (CH_3)_3O^+ \cdots \text{O-(Si,Al)}^- + CH_3OH$

methanol. The conjugate base of the acid site serves as the counterion for the trimethyloxonium cation. When care is take to ensure quantitation, we always observe a methanol signal (shoulder at 50 ppm) at ⅓ the integrated intensity of the trimethyloxonium resonance. Further evidence for the role of acid sites in Scheme II was the failure to produce any trimethyloxonium on either inactive NaZSM-5 or more weakly acidic zeolites such as HY. Referring again to Figure 8, further heating to 373 K results in more trimethyloxonium, but this decomposes at 473 K prior to hydrocarbon synthesis.

Is trimethyloxonium an intermediate in MTG chemistry? Several observations suggest that it may not be.

1. The oxonium ions are not formed in observable quantities in the presence of methanol concentrations typical of those in our experiments on methanol conversion (e. g., Figs 3 and 4). The oxonium ions are reasonably strong methylating agents, and excess methanol may shift the equilibrium back toward the ether.

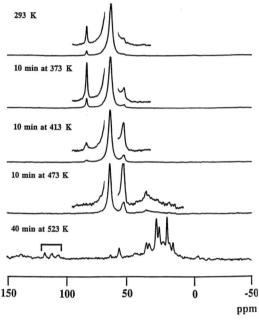

Figure 8 In situ ^{13}C CP/MAS NMR spectra showing the reactions of dimethyl ether-1-^{13}C on zeolite HZSM-5 as the sample was heated from 298 K to 523 K. Highlighted regions show the increase and decrease of trimethyloxonium upon heating. Bracket denotes background signals from Kel-F caps.

2. We have never observed spectroscopic signals suggestive of the oxonium ylide.

3. We have not observed the formation of dimethylethyloxonium or methyl ethyl ether from trimethyloxonium, even though the two ions are stable over the same temperature range.

4. As the temperature is raised, the concentration of trimethyloxonium decreases below the detection limit before hydrocarbon synthesis is observed.

It is certainly possible to come up with ad hoc arguments to explain away the above problems, but we are not satisfied with such arguments. In order to further explore the possible relevance of onium mechanisms in MTG chemistry, we decided to study a reactant analogous to dimethyl ether - dimethyl sulfide. The experience of synthetic chemists suggested that trimethylsulfonium should be more stable than trimethyloxonium. Sulfur is actually not a very good model for oxygen chemistry, but in this case the analogy had already been made. In a study of reactions on bifunctional catalysts, Olah and coworkers[25] proposed that dimethyl ether, dimethyl sulfide, and a variety of other heteroatom-substituted methane derivatives produced hydrocarbons through onium-ylide mechanisms analogous to Scheme I. ^{13}C labeled trimethylsulfonium formed readily in zeolite HZSM-5 when dimethyl sulfide and methanol-^{13}C were coadsorbed at room temperature in a glass ampule (Scheme III). The use of methanol-^{13}C eliminated the need for a custom synthesis of labeled dimethyl sulfide.

$$CH_3SCH_3 + CH_3OH + \underset{Si\ Al}{\overset{H}{\underset{|}{O}}} \longrightarrow \underset{\underset{Si\ Al}{\overset{\bar{O}}{|}}}{\overset{H_3C\ \ CH_3}{\underset{|}{\overset{\diagdown\ \diagup}{S^+}}}} + H_2O$$

Scheme III

Figure 9 shows ^{13}C MAS NMR spectra of this system. Methanol (51 ppm), dimethyl sulfide (17 ppm), and trimethylsulfonium (28 ppm) were present in the room temperature spectrum. Dimethyl sulfide was observed as a result of ^{13}C label exchange with trimethylsulfonium. Heating of the sample to 473 K resulted in the partial conversion of methanol to dimethyl ether (61 ppm). After the temperature was increased to 573 K, the only observable peak was from trimethylsulfonium. Even prolonged heating at 673 K resulted in little change in the trimethylsulfonium concentration. No peaks were observed which could

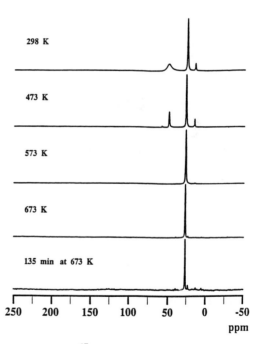

Figure 9 ^{13}C MAS NMR spectra of the reactions of an equimolar quantity of unlabeled dimethyl sulfide and methanol-^{13}C adsorbed on HZSM-5. ^{13}C-labeled trimethylsulfonium was readily formed at 298 K and remained stable at 673 K.

be assigned to the sulfonium ylide. This experiment indicates that trimethylsulfonium is very stable in the zeolite.

What do the trimethylsulfonium studies say about the oxonium mechanisms proposed for MTG chemistry? To the extent which the analogy between sulfur and oxygen is valid, these studies of sulfonium ions in HZSM-5 argue against oxonium mechanisms.

1. Trimethylsulfonium did not form hydrocarbons, even at 673 K.

2. We have not observed signals that can be assigned to sulfonium ylides.

3. We have not seen dimethylethylsulfonium form when MTG chemistry is carried out in the presence of trimethylsulfonium.

4. Even when dimethylethylsulfonium is generated directly in the zeolites, it does not eliminate ethylene at 573 K.

If trimethylsulfonium is a good analog for the less-easily studied oxonium ion, these investigations argue against oxonium routes. As this proceedings volume goes to press, the mechanism of MTG chemistry is still not known.

3 IN SITU ^{13}C MAS STUDIES OF METHANOL SYNTHESIS CHEMISTRY ON $Cu/ZnO/Al_2O_3$

The final application of in situ MAS NMR considers a non-zeolitic reaction. Essentially all of the world's methanol production relies on a ternary catalyst composed of copper, zinc oxide, and alumina (hereafter $Cu/ZnO/Al_2O_3$).[35-39] This catalyst converts synthesis gas to methanol with greater than 99% selectivity, a performance all the more remarkable considering the favorable thermodynamics for the formation of side products such as hydrocarbons, higher alcohols, and ethers. In spite of the near universal use of this catalyst and 25 years of study, there remains widespread disagreement about its structure and function. These issues are customarily introduced in the context of several questions:

1. Is methanol synthesized from CO or CO_2? Synthesis gas (which is obtained by steam reforming of natural gas) is a mixture of CO, CO_2 and H_2, and either carbon oxide can be reduced to methanol.

$$CO + 2 H_2 \rightleftharpoons CH_3OH \qquad (1)$$

$$CO_2 + 3 H_2 \rightleftharpoons CH_3OH + H_2O \qquad (2)$$

One reason why this most basic issue has proven difficult to settle is that $Cu/ZnO/Al_2O_3$ also catalyzes the water-gas shift reaction.

$$CO + H_2O \rightleftharpoons CO_2 + H_2 \qquad (3)$$

Furthermore, it is probable that more than one mechanism is operable, at least under certain conditions. Recently, Chinchen and coworkers at ICI have reported radiotracer studies that provide compelling evidence that methanol is synthesized from CO_2 under the conditions used in commercial reactors.[40]

2. What is the state of the copper in the working catalyst? Evidence exists for a variety of copper species, some of them dispersed in the ZnO phase. For example, Solomon and coworkers have used x-ray absorption edge and EXAFS measurements to study Cu/ZnO.[41] A significant amount of Cu(II) was found to be doped into the ZnO lattice in the calcined material. Following reduction, the catalyst contained metallic Cu, small Cu clusters, a Cu_2O phase and Cu(I) doped into the ZnO lattice. The ICI group has reported that about 30% of the initial copper metal surface is covered with oxygen under working conditions and that both the bare metal and oxide layer play important roles.[42]

3. What roles are played by ZnO and Al_2O_3? Before the development of $Cu/ZnO/Al_2O_3$, ZnO/Cr_2O_3 was used as the catalyst, but the temperature and pressure requirements were more severe. The current process is typically carried out at 523 K and 50-100 atm, whereas the ZnO/Cr_2O_3 catalyst required 623 K and 250-350 atm. Although a variety of roles have been proposed for ZnO in the current process, some workers assert that copper is the active catalyst (based on correlations between rate and metal surface area) and that the role of the oxides is to prevent sintering.[38,43] One criticism of that conclusion is that important, but not rate limiting, steps could still occur on ZnO.[44] The ZnO also deactivates the acid sites on Al_2O_3, preventing the synthesis of dimethyl ether.

4. What is the reaction mechanism? A wide variety of mechanisms have been proposed for this process.[35-39] A simplified scheme, favored by the ICI group is shown below.[38]

$$CO_2 \rightleftharpoons CO_{2\,ads}$$
$$H_2 \rightleftharpoons 2H_{ads}$$
$$CO_{2\,ads} + H_{ads} \rightleftharpoons HCOO_{ads}$$
$$HCOO_{ads} + 3H_{ads} \rightleftharpoons CH_3OH + O_{ads}$$
$$CO + O_{ads} \rightleftharpoons CO_2$$
$$H_2 + O_{ads} \rightleftharpoons H_2O \qquad (4)$$

Some of the above steps are probably composites of several elementary reactions. The existence of several intermediates, including formate, has been inferred based on temperature-programmed desorption studies.[45,46] Several infrared investigations of $Cu/ZnO/Al_2O_3$ and related catalysts have also been reported.[47-54]

Our approach has been to study the structure, dynamics and reactivity of a the reactants, intermediates and products obtained by adsorption of CH_3OH, $HCOOH$, H_2CO, CO_2 and CO. In order to further elucidate the roles of the catalyst components, the reactions of some or all of the above species were studied on a variety of catalysts: $Cu/ZnO/Al_2O_3$, ZnO/Al_2O_3, ZnO, Al_2O_3 and Cu/Al_2O_3. The reactions of these catalyst/adsorbate samples were monitored in situ by obtaining ^{13}C magic-angle spinning (MAS) NMR data as the sealed samples were heated in the NMR probe.

The structure, dynamics, and reactivity of species formed after the adsorption of methanol-^{13}C, formaldehyde-^{13}C, and formic acid-^{13}C on $Cu/ZnO/Al_2O_3$, ZnO/Al_2O_3, Al_2O_3, and ZnO were studied in detail. Some experiments were also performed on Cu/Al_2O_3, but these were limited due to sintering during preparation or subsequent treatment. Following adsorption at room temperature, the principal components of the ^{13}C chemical shift tensor of the adsorbate were measured, and in some cases detailed relaxation measurements were performed. The samples were then heated in the NMR probe. Several general protocols were used. In some cases, the sample temperature was raised incrementally, and spectra were obtained at each temperature and after cooling to ambient. In other cases the sample was heated directly to the maximum temperature of 523 K, and spectra were obtained over a time course and after cooling to room temperature.

Methanol readily adsorbed on all of the catalysts, and its properties in the adsorbed state were characterized in detail as a probe of surface properties and as a prelude to in situ studies of its decomposition. Figure 10 presents ^{13}C CP/MAS spectra obtained after the adsorption of methanol-^{13}C on the catalysts at room temperature. The appearance of these spectra is summarized as follows: (1) a ^{13}C resonance with an isotropic chemical shift of 49 ppm was observed in spectra of all catalysts; (2) on catalysts which contained ZnO, a second isotropic peak was observed at 54 ppm that was accompanied by spinning sidebands; (3) both the 49- and 54-ppm peaks survived 50 μs of interrupted decoupling; and (4) spinning sidebands from the 49-ppm resonance were observed for all catalysts containing Al_2O_3. The presence of spinning sidebands is characteristic of species bound in such a fashion that they are incapable of isotropic molecular motion on a timescale of ca. 1 ms. Hence, the sideband pattern reflects either the chemical shift anisotropy (CSA) of the nuclear site or the residual CSA after averaging by restricted molecular motion.

From detailed relaxation studies and measurements of the principal components of ^{13}C chemical shift tensors (vide infra) the following picture of the adsorption of methanol on these catalysts emerged: Methanol adsorbs on Al_2O_3 as two distinct species, one tightly bound (chemisorbed) and the other highly mobile and easily removed by evacuation (physisorbed). The 54-ppm peak which forms on catalysts containing ZnO is due to a second species covalently bound to the surface. Intact methanol also adsorbs on ZnO, but it is easily removed. We did not observe Knight shifts in any of the ^{13}C spectra of adsorbates on Cu/Al_2O_3 or $Cu/ZnO/Al_2O_3$; but on the latter catalyst, adsorbed methanol exhibited relaxation behavior consistent with interaction with paramagnetic copper sites (either dispersed Cu(0) or Cu(II)).

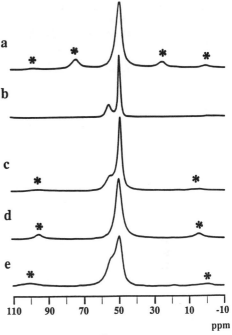

Figure 10. ^{13}C CP/MAS spectra obtained after the adsorption of methanol-^{13}C at room temperature on: (a) Al_2O_3; (b) ZnO; (c) ZnO/Al_2O_3; (d) Cu/Al_2O_3; and (e) $Cu/ZnO/Al_2O_3$. * denotes spinning sidebands.

Table 2 shows ^{13}C and ^1H relaxation data for methanol adsorbed on the catalysts. ^1H and ^{13}C $T_{1\rho}$ values do not appear to be diagnostic of either the catalyst surface or the structure of the adsorbate. ^1H T_1 values are approximately an order of magnitude lower for catalysts containing copper. The ^{13}C T_1 results are most informative. On Al_2O_3, the 49-ppm methanol peak exhibits biphasic relaxation, and only the long component persists after evacuation to remove physisorbed methanol. The chemisorbed methanol species with a long ^{13}C T_1 also displays significant spinning sidebands in a slow speed MAS spectrum (Figure 11). Infrared studies have previously shown that methoxy groups form on ZnO.[55] The surface-bound methoxy signal (54 ppm) has a longer ^{13}C T_1 than the more mobile physisorbed methanol (49 ppm). The latter signal is readily removed by evacuation (Figure 12) to leave almost exclusively the surface-bound methoxy species which displayed the prominent sideband patterns characteristic of restricted reorientation.

Table 2 Relaxation Times[1] for the Species Formed after the Adsorption of Methanol-^{13}C on Various Catalysts at 298 K.

Catalyst	Peak[2]	$^{13}C\ T_1$	$^{13}C\ T_{1\rho}$	$^1H\ T_1$	$^1H\ T_{1\rho}$
Al_2O_3[3]	49	418 1389	NP	NP	NP
Al_2O_3[4]	49	2778	27	NP	93
ZnO[3]	49	955	19	111	NP
	54	1636	31	129	
ZnO/Al_2O_3[3]	49	1390	17	187	88
	54	2190	44	456	90
$Cu/ZnO/Al_2O_3$[3]	49	181 1226	26	15	87
	54	298 1751	46	17	88
$Cu/ZnO/Al_2O_3$[4]	49	331 1832	46	NP	NP
	54	464 2223	63		
$Cu/ZnO/Al_2O_3$[5]	49	112 1377	NP	NP	NP
	54	284 2206			
Cu/Al_2O_3[3]	49	508	20	15	86

[1]All in ms, when two times are given for a single peak, a biexponential decay of magnetization is implied.
[2]In ppm relative to TMS.
[3]Sample not evacuated after adsorption of methanol.
[4]Sample evacuated after adsorption to remove physisorbed methanol.
[5]20% $^{13}CH_3OH$, 80% $^{12}CH_3OH$ adsorbed followed by evacuation.
NP denotes measurement not performed.

All of the $Cu/ZnO/Al_2O_3$ samples studied displayed biphasic $^{13}C\ T_1$ behavior for both the 49- and 54-ppm signals. Although the exact values of the relaxation rates varied from sample to sample, the fast and slow components differed by a factor of 4-10 in all cases. We also studied the relaxation behavior of 20% ^{13}C enriched methanol on $Cu/ZnO/Al_2O_3$ to guard against the possibility that $^{13}C-^{13}C$ spin diffusion was obscuring even more complicated relaxation behavior, but the biphasic $^{13}C\ T_1$ was also observed with that sample (Table 2). The most plausible explanation for the short component is relaxation induced by paramagnetic copper sites dispersed in an oxide phase. Since the catalyst samples were reduced, these are probably Cu(0) sites.

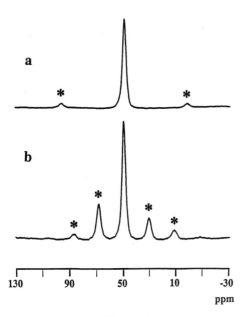

Figure 11 ^{13}C CP/MAS spectra acquired after the adsorption of methanol-^{13}C on Al$_2$O$_3$ and evacuation to remove excess methanol (a) at a spinning speed of 3.49 kHz and (b) 1.39 kHz. * denotes spinning sidebands.

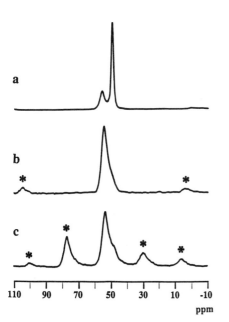

The experiments described above provide evidence for the formation of two chemisorbed species upon exposure of Cu/ZnO/Al$_2$O$_3$ catalyst to methanol. One is associated with Al$_2$O$_3$ and the other with ZnO. The properties of these species were further elucidated using slow spinning spectra of the type shown in Figures 11 and 12, and calculation of the principal components of the ^{13}C chemical shift tensors using the method of Herzfeld and Berger.[56] Those results are compared with literature data for appropriate model compounds in Table 3. The $\Delta\delta$ value for chemisorbed methanol on Al$_2$O$_3$ (49 ppm) is slightly smaller than for the model compounds (53-63 ppm). We were able to account for the observed chemical shift parameters of chemisorbed methanol by operating on the chemical shift tensor of solid methanol with the rotation implied in (**V**). Excellent agreement between experiment and calculation was ob-

Figure 12 ^{13}C CP/MAS spectra obtained (a) after the adsoption of methanol-^{13}C on ZnO; (b) after the adsorption of methanol-^{13}C on a shallow bed of ZnO and evacuation to remove physisorbed methanol; and (c) for sample in (b) at a spinning speed of 1.63 kHz. * denotes spinning sidebands

tained for Al-O-C bond angles of 70°.

In the case of the surface-bound methoxy species on ZnO, the chemical shift anisotropy ($\Delta\delta$) is significantly

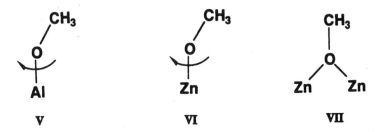

V   VI   VII

larger than the values of all of the model compounds. Clearly, the difference in the chemical shift tensor for the surface-bound methoxy group cannot be reconciled with

Table 3 ^{13}C Chemical Shift Parameters[1] for Surface-Bound Methoxy Species and Suitable Model Compounds.

Sample	T (K)	δ_{iso}	δ_{11}	δ_{22}	δ_{33}	$\Delta\delta$	η	ref
Surface-bound methoxy on Al_2O_3	298 K	49	65	65	17	40	0	This work
Surface-bound methoxy on ZnO	298 K	54	79	79	5	74	0	This work
Surface-bound methoxy on ZnO	273 K	54	79	79	4	75	0	This work
Surface-bound methoxy on ZnO	233 K	53	79	79	2	77	0	This work
Surface-bound methoxy on ZnO	193 K	53	80	80	0	80	0	This work
CH_3OH	177 K	52	73	73	10	63	0	57
p-dimethoxy-benzene	298 K	56	80	72	16	60	0.12	58
$KOCH_3$	298 K	58	78	78	18	60	0	58
$Mg(OCH_3)_2$	298 K	53	74	74	10	64	0	57
$Ca(OCH_3)_2$	298 K	55	73	73	20	53	0	57

[1] δ's and $\Delta\delta$'s are in ppm relative to TMS.

those for the model compounds by invoking motional averaging; an electronic (structural) difference is required. Two structures could be imagined for a surface-bound methoxy species, terminal (**VI**) and bridging (**VII**). The terminal form would presumably be capable of rotation about the Zn-O axis, and that would lead to a relatively small value for $\Delta\delta$ (c.f. species **V**). In contrast, the bridging methoxy would only be capable of motion about the O-C bond axis. In a study of methanol adsorption on zeolite HZSM-5, Dybowski and coworkers[28] characterized a terminal methoxy species analogous to **V**. The ^{13}C T_1 behavior of that species was governed by rotation about the Al-O bond with an with activation energy of 11 kcal/mol. We performed variable temperature ^{13}C T_1 measurements for the methoxy group on ZnO and determined that the motion responsible for T_1 was characterized by an activation energy of 2.2 kcal/mol -- exactly what one would expect for unrestricted methyl rotation. The bridging methoxy, **VII**, satisfactorily accounts for the dynamics which we observed. Furthermore, it is structurally different from the model compounds, so differences in the principal components of the chemical shift tensor as well as isotropic chemical shift are not unreasonable. One possible explanation for the formation of bridging methoxy species is that methanol reacts with an oxygen vacancy on the ZnO surface.

Having characterized the adsorption of methanol on $Cu/ZnO/Al_2O_3$ and related catalysts in detail, we then studied its decomposition using in situ MAS NMR. A typical in situ run involved the acquisition of several dozen to several hundred ^{13}C MAS spectra of the species formed at high temperature probing the structure and the relaxation behavior of those species over a range of temperatures. Several ^{13}C MAS spectra from a representative experiment are shown in Figure 13. Chemisorbed methanol and surface-bound methoxy originally present on $Cu/ZnO/Al_2O_3$ at 298 K (Figure 13a) began forming a formate species immediately after increasing the sample temperature to 353 K (Figure 13b). Considerably more formate formed after the temperature was raised to 453 K (Figure 13c). Upon raising the sample temperature to 523 K, the formate intermediate was converted into a carbonate (or bicarbonate) intermediate which has spectroscopic properties very distinct from the formate intermediate in spite of similar ^{13}C isotropic chemical shifts (vide infra). Figs 13d and e show two spectra of this sample which were obtained immediately after cycling back to 298 K. The cross polarization spectrum (13d) shows an intense signal for the carbonate intermediate rigidly bound to the catalyst surface, whereas the Bloch decay spectrum (13e) not only shows the carbonate, but also reveals that approximately 15% of the carbonate had decomposed to gas phase or physisorbed CO_2. The above experiment then shows the decomposition of chemisorbed methanol and surface-bound methoxy to CO_2, by way of surface-bound formate and carbonate.

Analogous in situ experiments were also performed on Al_2O_3, ZnO/Al_2O_3, and ZnO in order to further elucidate the roles of the catalyst components. None of these catalysts were active for methanol oxidation at the temperature studied, and we conclude that the copper catalyzes the conversion of methoxy to formate. Dimethyl ether formed on the Al_2O_3 catalyst, but not on the ZnO/Al_2O_3 catalyst. Indeed, one of the known roles of ZnO is to deactivate the acidic sites on Al_2O_3 and prevent dimethyl ether formation.

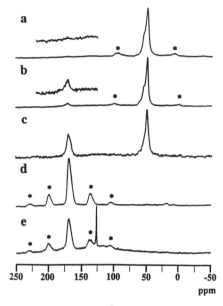

Figure 13 ^{13}C MAS spectra showing in situ decomposition of methanol-^{13}C on Cu/ZnO/Al$_2$O$_3$: (a) CP spectrum at room temperature; (b) CP spectrum at 353 K; (c) CP spectrum at 453 K; (d) CP spectrum obtained after heating to 523 K and cooling to room temperature; and (e) Bloch decay spectrum obtained after heating to 523 K and cooling to room temperature.

All of the catalysts readily adsorbed formic acid at room temperature to form a surface-bound formate species. ^{13}C chemical shift data for formate species were obtained by Herzfeld-Berger analysis for all of the samples studied. These data are summarized in Table 4. The formate species obtained by adsorption of formic acid had essentially identical chemical shift parameters on all of the catalysts.

Three bonding modes are possible for formate species on surfaces, monodentate **VIII**, bidentate **IX**, and bridging **X**. Examination of the shielding tensor data in Table 4

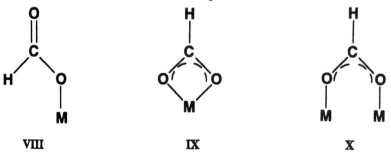

suggests that the formates formed from direct adsorption

Table 4 ^{13}C Chemical Shift Parameters[1] for Surface-Bound Formates and Suitable Model Compounds.

Sample	T (K)	δ_{iso}	δ_{11}	δ_{22}	δ_{33}	$\Delta\delta$	η	ref
Formic Acid on Al$_2$O$_3$	298 K	170	232	173	106	96	0.92	This work
Formic Acid on ZnO	298 K	171	234	175	105	99	0.89	This work
Formic Acid on ZnO/Al$_2$O$_3$	298 K	172	235	177	105	100	0.87	This work
Formic Acid on ZnO/Al$_2$O$_3$	493 K	172	217	173	127	68	0.98	This work
Formic Acid on Cu/Al$_2$O$_3$	298 K	172	238	171	108	99	0.95	This work
Formic Acid on Cu/ZnO/-Al$_2$O$_3$	298 K	172	235	176	106	99	0.89	This work
Methanol Decomposition on Cu/ZnO/-Al$_2$O$_3$	298 K	172	226	164	120	84	0.79	This work
Paraformaldehyde Disproportionation on Cu/ZnO/Al$_2$O$_3$	298 K	172	224	166	125	78	0.79	This work
Formic Acid	125 K	168	251	162	92	125	0.84	59
Ammonium Formate	185 K	164	225	168	99	98	0.88	59
Calcium Formate	298 K	173	231	184	105	102	0.69	60

[1] δ's and $\Delta\delta$'s are in ppm relative to TMS.

of formic acid at room temperature are not capable of large amplitude motion; hence the bidentate or bridging species seem most probable. This conclusion is supported by an independent line of reasoning. In a ^{13}C solid-state NMR study of formic acid adsorbed in Y zeolites,[59] Duncan and Vaughan noted that there is a good correlation between δ_{22} and the ratio of the C-O bond lengths of formate species, i.e., δ_{22} shifts from the upfield part of the powder pattern to the downfield part as the C-O bond lengths become more symmetric. This correlation is exhibited by the ratio of the disparity between the upfield and central components ($\delta_{22}-\delta_{33}$) to the overall anisotropy ($\delta_{11}-\delta_{33}$). For methyl formate, where the C-O bond lengths are not the same, this ratio is equal to 0.2. For calcium formate, where the formate ion is chelated to two calcium ions and the C-O bond lengths are the same, the ratio is equal to 0.6. Duncan and Vaughan suggested that, in general, $(\delta_{22}-\delta_{33})/(\delta_{11}-\delta_{33})$ is on the order of 0.6 for symmetrical species, and 0.2 for ester-like formate species. For the surface-bound formate species formed from the adsorption of formic acid on Cu/ZnO/Al$_2$O$_3$ and related catalysts (see Table 5), $(\delta_{22}-\delta_{33})/(\delta_{11}-\delta_{33})$ ranged from 0.48 to 0.55, and we conclude that these species are best described as bidentate or bridging.

The formate species formed by the decomposition of methanol on Cu/ZnO/Al$_2$O$_3$ had chemical shift parameters that were slightly different from those measured for species formed from the direct adsorption of formic acid. The $\Delta\delta$ value was approximately 15% lower, but the value of the asymmetry parameter η and the ratio $(\delta_{22}-\delta_{33})/(\delta_{11}-\delta_{33})$ were consistent with the values for the other species. It is not clear whether the slightly smaller value of $\Delta\delta$ reflects an electronic difference or restricted molecular motion. In FT-IR studies of formate species on a variety of metal oxides, Busca and coworkers[52] noted that the infrared frequencies of surface formates were different depending on whether the formate was obtained

Table 5 ^{13}C Chemical Shift Parameters[1] for Surface-Bound Carbonates and Suitable Model Compounds.

Sample	T (K)	δ_{iso}	δ_{11}	δ_{22}	δ_{33}	$\Delta\delta$	η	ref
CO$_2$ on Cu/ZnO/Al$_2$O$_3$	298 K	166	225	155	115	89	0.68	This work
Decomposition of Formic Acid on Cu/ZnO/Al$_2$O$_3$	298 K	166	229	161	114	95	0.75	This work
Decomposition of Methanol on Cu/ZnO/Al$_2$O$_3$	298 K	166	235	177	114	104	0.91	This work
Calcium Carbonate	493 K	169	217	173	119	75	0.88	61
Barium Carbonate	298 K	170	238	171	118	102	0.78	62
Potassium Bicarbonate	298 K	162	218	155	113	84	0.75	62
Ammonium Bicarbonate	298 K	163	223	152	119	90	0.55	62

[1] δ's and $\Delta\delta$'s are in ppm relative to TMS.

by adsorption of formic acid or oxidation of methanol.

Upon heating the formic acid/Cu/ZnO/Al$_2$O$_3$ sample to 523 K, formation of carbonate and CO$_2$ was observed as noted previously for the methanol/Cu/ZnO/Al$_2$O$_3$ system. The formic acid/ZnO/Al$_2$O$_3$ sample was much less reactive, and only a trace of carbonate was obtained after heating the MAS rotor to 523 K. This result again underscores the role of copper in redox reactions on the catalyst. The low reactivity of formate on ZnO/Al$_2$O$_3$ did allow us to determine that formate is mobile on the catalyst at elevated temperatures. Upon heating from 298 K to 493 K (Table 4), the $\Delta\delta$ decreased from 102 ppm to 68 ppm, a result consistent with large amplitude molecular motion. We conclude that formate is mobile on the catalyst surface at reaction temperature.

In spite of their similar isotropic ^{13}C chemical shifts, formate and carbonate could always be distinguished based on their response to interrupted decoupling. The formate signals formed by adsorption of formic acid or the decomposition of methanol or paraformaldehyde at ca. 373 K gave ^{13}C signals that did not survive interrupted decoupling, as expected for a protonated carbon in a relatively rigid species. Signals due to carbonate species were not significantly attenuated by interrupted decoupling.

The principal components of the ^{13}C chemical shift tensors were also characterized for carbonate species formed on some of the catalysts studied; these data are reported in Table 5. Inspecting that Table, one notes reasonable agreements between principal component data of surface species and carbonate or bicarbonate model compounds.

Several investigations have suggested an intermediary role of formaldehyde in methanol synthesis.[47,48,51,52,60] For example, Madix and coworkers have proposed that methanol decomposition proceeds by the copper dehydrogenating a methoxy group to formaldehyde which is then oxidized to formate.[60] We have never observed an NMR signal attributable to formaldehyde, or formaldehyde complexes, in in situ studies of methanol decomposition. This does not rule out a role for formaldehyde if one assumes that the formaldehyde is short-lived under the reaction conditions used in this investigation. We investigated the chemistry of formaldehyde on Cu/ZnO/Al$_2$O$_3$ and the other catalysts by decomposing paraformaldehyde-^{13}C on the samples in in situ NMR experiments at 373 K. Paraformaldehyde decomposed on all of the catalysts, including pure Al$_2$O$_3$ and pure ZnO, to give equal moles of surface-bound methoxy groups and formate. This result is reminiscent of the Cannizzarro disproportionation of formaldehyde.[48,54] Representative spectra are reported in Figure 14. Free formaldehyde was not observed in any of the reactions. Formate and methoxy were converted to carbonate (e.g.

Figure 14c) upon heating the copper-containing catalyst to high temperature, but catalysts without copper were unreactive.

Figure 14d shows the ^{13}C spectrum of paraformaldehyde-^{13}C on ZnO/Al$_2$O$_3$ at room temperature after heating to 373 K. In addition to the signals for formate and methoxy groups and/or methanol, there are two smaller isotropic resonances: a signal at 89 ppm due to unreacted paraformaldehyde, and a signal at 96 ppm. Neither signal survives interrupted decoupling. One mode in which formaldehyde bonds to surfaces is by the formation of the dioxomethylene species **XI**.[52-54] We are unable to unambiguously assign the 96-ppm signal, but species **XI** is a plausible assignment. Further heating of the ZnO/Al$_2$O$_3$ resulted in a decrease in both paraformaldehyde and the 96-ppm resonance (Figure 14e).

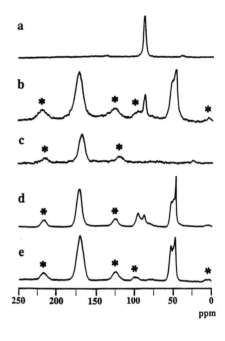

Figure 14 ^{13}C CP/MAS spectra showing the decomposition of paraformaldehyde-^{13}C on Cu/ZnO/Al$_2$O$_3$ (a - c) and ZnO/Al$_2$O$_3$ (d - e): (a) at 298 K; (b) after heating to 373 K and cooling to 298 K; (c) after heating to 523 K and cooling to 298 K; (d) the same as in (b); and (e) the same as in c. * denotes spinning sidebands.

The paraformaldehyde decomposition studies are consistent with an intermediary role for formaldehyde in methanol decomposition. We did not observe free formaldehyde in the methanol decomposition studies, but it was not observed in the paraformaldehyde decomposition reactions either. Formaldehyde is very reactive at 373 K for all of the catalysts studied and immediately undergoes oxidation or disproportionation, possibly through a dioxymethylene intermediate.

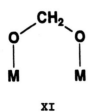

XI

Detailed relaxation measurements were performed on the paraformaldehyde-^{13}C catalyst samples at room temperature after heating to 373 K (not shown). The relaxation times for the methoxy carbon signals at 49 and 54 ppm followed the same trends as those seen previously for these signals in samples prepared by direct adsorption of methanol. Analogous trends were exhibited by the relaxation data for the formate signals. Very similar formate relaxation data were measured for samples prepared by direct adsorption of formic acid (not shown). As seen previously in Table 2, the ^{13}C T_1 values were most diagnostic of catalyst composition. Biexponential relaxation was observed for all three species of the $Cu/ZnO/Al_2O_3$ catalyst -- again suggestive of Cu(0) dispersed into the oxide phase. The similarities in relaxation times observed for direct adsorption vs. in situ synthesis suggests that these are in fact the same species.

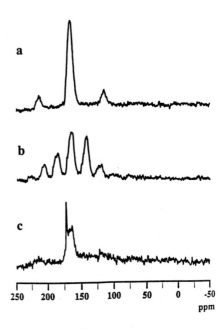

Figure 15 ^{13}C CP/MAS spectra probing the adsorption $^{13}CO_2$ and ^{13}CO on $Cu/ZnO/Al_2O_3$ at 298 K: (a) after adsorption of $^{13}CO_2$, spinning speed 3.77 kHz; (b) same as above except spinning speed 1.60 kHz; (c) after adsorption of ^{13}CO, spinning speed ca. 3.8 kHz.

The requirement of 50-100 atm for converting CO_2 or CO and H_2 to methanol precludes the study of the forward reaction by magic-angle spinning NMR with currently available rotor technology. Nevertheless, some results were obtainable for the first step of the reactions of CO_2 and CO on $Cu/ZnO/Al_2O_3$ at low pressure. CO_2 was easily taken up by the catalyst to form a carbonate (Figure 15) with ^{13}C chemical shift data similar to those for carbonates formed from methanol or formic acid decomposition (Table 5). Two signals were observed when ^{13}CO was adsorbed on $Cu/ZnO/Al_2O_3$ at low pressure (Figure 15c). The broad signal at 166 ppm is due to carbonate, but the assignment of the very narrow resonance at 172 ppm is more speculative. Although that chemical shift is identical to that of the formate species observed in this contribution, the signal survived interrupted decoupling and otherwise appeared to be a mobile species. Gas phase CO has a ^{13}C chemical shift of 184 ppm, and it usually shifts downfield upon forming strong metal carbonyl complexes. In the case of weak complexes, however, upfield shifts are more common. Cu(I) carbonyl compounds[61-63] are among the weakest carbonyl complexes ever isolated, and the ^{13}C chemical shifts are in the vicinity of 172 ppm.[64] Zn carbonyls have not been investigated by NMR, but they are also expected to involve weak bonding.[65-67] We speculate that the 172-ppm resonance in Figure 15c is a mobile CO on the surface weakly interacting with either copper or zinc.

The in situ NMR studies presented here enable us to comment on the questions posed earlier:

1. Both CO_2 and CO interact with the catalyst surface at low pressure, but CO_2 is taken up more readily to form the carbonate species observed during methanol decomposition on $Cu/ZnO/Al_2O_3$. CO_2 is also formed when the carbonate species formed from methanol is heated to 523 K.

2. ^{13}C T_1 relaxation time measurements provide indirect evidence that paramagnetic Cu(0) atoms or clusters are dispersed into the oxide phase(s) in $Cu/ZnO/Al_2O_3$. This result does not disprove the claim that only the metal surface is important, but the possibility of a role for dispersed clusters and/or a synergistic effect between the metal and oxide phases should not be overlooked.

3. Pure ZnO, pure Al_2O_3, and ZnO/Al_2O_3 were shown to be fairly reactive in this contribution. A terminal methoxy group forms on Al_2O_3 and a bridging methoxy forms on ZnO. Formaldehyde readily disproportionated to formate and methoxy on any of the above catalysts. Copper was essential for the oxidation of formate to carbonate and the oxidation of methoxy to formaldehyde and/or formate at 523 K. Methanol synthesis clearly cannot proceed at 523 K without a metal component, but the oxide phases appear to be more reactive than necessary for a mere support role. Copper alone may be capable of catalyzing all of the steps in methanol synthesis; but under the in situ conditions used in this investigation, several adsorbed species are present on the ZnO and Al_2O_3 components.

4. The most significant results of this application pertain to the reaction mechanism. The terminal bridging methoxy on Al_2O_3 and the bridging methoxy on ZnO form from either formaldehyde decomposition or methanol adsorption. Formaldehyde is too reactive to be detected as a free species using our in situ protocol, but the paraformaldehyde decomposition studies are consistent with formaldehyde having an intermediary role -- possibly as a surface-bound dioxymethylene species. Formate is a ubiquitous intermediate in many catalysis reactions, and it was easily generated and thoroughly characterized in the present investigation. The formate on $Cu/ZnO/Al_2O_3$ and related catalysts is best described as a symmetrical bidentate or bridging species which is rigid at 298 K but exhibits anisotropic motion on the surface on a time scale of ca. 1 ms at 493 K. Surface carbonate is produced readily by oxidation of formate on $Cu/Zn/Al_2O_3$, and the carbonate is in equilibrium with free CO_2.

4 ACKNOWLEDGEMENT

Some of the people whose work is described in this

chapter include Eric Munson, Jeff White, Larry Beck, and Noel Lazo.

This work was supported by the National Science Foundation (Grant CHE-8918741).

REFERENCES

1. J. F. Haw, B. R. Richardson, I. S. Oshiro, N. D. Lazo and J. A. Speed, J. Am. Chem. Soc., 1989, 111, 2052.
2. B. R. Richardson, N. D. Lazo, P. D. Schettler, J. L. White and J. F. Haw, J. Am. Chem. Soc., 1990, 112, 2886.
3. N. D. Lazo, B. R. Richardson, P. D. Schettler, J. L. White, E. J. Munson and J. F. Haw, J. Phys. Chem., 1991, 95, 9420.
4. N. D. Lazo, J. L. White, E. J. Munson, M. Lambregts and J. F. Haw, J. Am. Chem. Soc., 1990, 112, 4050.
5. M. J. Lambregts, E. J. Munson, A. A. Kheir and J. F. Haw, J. Am. Chem. Soc., 1992, in press.
6. J. L. White, N. D. Lazo, B. R. Richardson and J. F. Haw, J. Catal., 1990, 125, 260.
7. F. G. Oliver, E. J. Munson and J. F. Haw, J. Phys. Chem., 1992, in press.
8. E. J. Munson, N. D. Lazo, M. E. Moellenhoff and J. F. Haw, J. Am. Chem. Soc., 1991, 113, 2783.
9. E. J. Munson and J. F. Haw, J. Am. Chem. Soc., 1991, 113, 6303.
10. N. D. Lazo, D. K. Murray, M. L. Kieke and J. F. Haw, J. Am. Chem. Soc., 1992, in press.
11. J. L. White, L. W. Beck and J. F. Haw, J. Am. Chem. Soc., 1992, in press.
12. J. L. White, L. W. Beck and J. F. Haw, 1992, submitted.
13. L. W. Beck, J. L. White and J. F. Haw, J. Magn. Reson., 1992, in press.
14. J. F. Haw and J. A. Speed, J. Magn. Reson., 1988, 78, 344.
15. E. J. Munson, D. B. Ferguson, A. A. Kheir and J. F. Haw, J. Catal., 1992, in press.
16. D. B. Venuto and E. T. Habib, 'Fluid Catalytic Cracking with Zeolite Catalysts', Marcel Dekker, New York, 1979, p. 124.
17. R. N. Young, Prog. NMR Spectrosc., 1979, 12, 261.
18. J. C. Rees and D. Whittaker, J. Chem. Soc. Perkin II, 1980, 948.
19. N. C. Deno, 'Carbonium Ions' (G. A. Olah and P. V. R. Schleyer, eds.), Wiley-Interscience, New York, 1970, Vol. 2, Chapter 18.
20. C. D. Chang, Catal. Rev. - Sci. Eng., 1983, 25, 1.
21. C. D. Chang, 'Perspectives in Molecular Sieve Science' (W. E. Flank and T. E. Whyte, eds.), ACS Symp. Ser. 368, 1988, p. 596.
22. C. D. Chang and A. J. Silvestri, J. Catal., 1977, 47, 249.

23. C. S. Lee and M. M. Wu, J. Chem. Soc, Chem. Commun., 1985, 250.
24. Y. Ono and T. J. Mori, J. Chem. Soc., Faraday Trans., 1981, 77, 2209.
25. G. A. Olah, A. Doggweiler, J. D. Feldberg, S. Frohlich, M. J. Grdina, R. Karpeles, T. Keumi, S. Inabe, W. M. Ip, K. Lammertsma, G. Salem and D. C. Tabor, J. Am. Chem. Soc., 1984, 106, 2143.
26. J. P. van den Berg, J. P. Wolthuizen and J. H. C. van Hooff 'Proceedings of the 5th International Conference on Zeolites' (L. V. Rees, ed.), Heyden, London, 1980.
27. J. K. Clarke, R. Darcy, B. F. Hegarty, E. O'Donoghue, V. Amir-Ebrahimi and J. R. Rooney, J. Chem. Soc., Chem. Commun. 1986, 425.
28. C. Tsiao, D. R. Corbin and C. Dybowski, J. Am. Chem. Soc., 1990, 112, 7140.
29. M. W. Anderson and J. Klinowski, J. Chem. Soc., Chem. Commun., 1990, 918.
30. J. E. Jackson and F. M. Bertsch, J. Am. Chem. Soc., 1990, 112, 9085.
31. M. W. Anderson and J. Klinowski J. Am. Chem. Soc., 1990, 112, 10.
32. M. W. Anderson and J. Klinowski, Nature, 1989, 339, 200.
33. G. O. Brunner, Zeolites, 1987, 7, 9.
34. G. J. Hutchings and P. Johnston, Appl. Catal., 1990, 57, L5.
35. H. H. Kung, Catal. Rev. - Sci. Eng., 1980, 22, 235.
36. J. C. J. Bart and R. P. A. Sneeden, Catal. Today, 1987, 2, 1.
37. G. C. Chinchen, P. J. Denny, J. R. Jennings, M. S. Spencer and K. C. Waugh, Appl. Catal., 1988, 36, 1.
38. G. C. Chinchen, K. Mansfield and M. S. Spencer, Chemtech, 1990, 20, 692.
39. K. Klier, Adv. Catal., 1982, 31, 243.
40. S. Kinnaird, G. Webb and G. C. Chinchen, J. Chem. Soc., Faraday Trans., 1987, 83, 3399.
41. L.-S. Kau, K. O. Hudgson and E. I. Solomon, J. Am. Chem Soc., 1989, 111, 7103.
42. M. Bowker, R. A. Hadden, H. Haughton, J. N. K. Hyland and K. C. Waugh, J. Catal., 1988, 109, 263.
43. G. C. Chinchen, P. J. Denny, D. G. Parker, G. D. Short, M. S. Spencer, K. C. Waugh and D. A. Whan, 'ACS Symposium on Methanol Synthesis', Fuel Chemistry Division American Chemical Society, Washington, DC, 1984.
44. G. C. Chinchen, K. C. Waugh and D. A. Whan, Appl. Catal., 1986, 25, 101.
45. C. Chauvin, J. Saussey, J.-C. Lavalley, H. Idriss, J. P. Hindermann, A. Kiennemann, P. Chaumette, Ph. Courty, J. Catal., 1990, 121, 56.
46. A. Kiennemann, H. Idriss, J. P. Hindermann, J.-C. Lavalley, A. Vallet, P. Chaumette and Ph. Courty, Appl. Catal., 1990, 59, 165.
47. J. F. Edwards and G. L. Schrader, J. Phys. Chem., 1984, 88, 5620.
48. J. F. Edwards and G. L. Schrader, J. Phys. Chem., 1985, 89, 782.

49. F. Boccuzzi, G. Ghiotti and A. Chiorino, Surf. Sci., 1985, 162, 361.
50. G. Ghiotti, F. Boccuzzi and A. Chiorino, 'Adsorption and Catalysis on Oxide Surfaces' (M. Che and G. C. Bond, eds.), Elsevier, Amsterdam, 1985.
51. J. F. Edwards and G. L. Schrader, J. Catal., 1985, 94, 175.
52. G. Busca, J. Lamotte, J.-C. Lavalley and V. Lorenzelli, J. Am. Chem Soc., 1987, 109, 5197.
53. H. Idriss, J. P. Hindermann, A. Kieffer, A. Kiennemann, A. Vallet, C. Chauvin, J.-C. Lavalley and P. Chaumette, J. Mol. Catal., 1987, 42, 205.
54. C. Li, K. Domen, K.-I. Maruga and T. Onishi, J. Catal., 1990, 125, 445.
55. D. L. Roberts and G. L. Griffin, J. Catal., 1986, 101, 201.
56. J. Herzfeld and A. E. Berger, J. Chem. Phys., 1980, 73, 6021.
57. I. D. Gay, J. Phys. Chem., 1980, 84, 3230.
58. M. M. Maricq and J. S. Waugh, J. Chem. Phys., 1979, 70, 3300.
59. T. M. Duncan and R. W. Vaughan, J. Catal., 1981, 67, 49.
60. I. E. Wachs and R. J. Madix, J. Catal., 1978, 53, 208.
61. M. Pasquali, C. Floriani, G. Venturi, A. Gaetani-Manfredotti and A. Chiesi-Villa, J. Am. Chem. Soc., 1982, 104, 4092.
62. M. Pasquali, C. Floriani and A. Gaetani-Manfredotti, Inorg. Chem., 1981, 20, 3382.
63. M. Pasquali, F. Marchetti and C. Floriani, Inorg. Chem., 1978, 17, 1684.
64. Y. Souoma, J. Iyoda and H. Sano, Inorg. Chem., 1976, 15, 968.
65. R. R. Gay, M. H. Nodine, V. E. Henrich, H. J. Zeiger and E. I. Solomon, J. Am. Chem. Soc., 1980, 102, 6752.
66. J. Lin, P. Jones, J. Gruckert and E. I. Solomon, J. Am. Chem. Soc., 1991, 113, 8312.
67. G. Ghiotti, F. Boccuzzi and A. Chiorino, J. Chem. Soc., Chem. Commun., 1985, 1012.

High Thermal Stability of Nickel–Aluminium Mixed Oxides Obtained from Hydrotalcite-type Precursors

P. Beccat[1], J. C. Roussel[1], O. Clause[1], A. Vaccari[2], and F. Trifiro[2]

[1] INSTITUT FRANÇAIS DU PÉTROLE, I & 4, AVENUE DE BOIS-PRÉAU, BP 311, 92506 RUEIL-MALMAISON CEDEX, FRANCE
[2] DEPARTMENT OF INDUSTRIAL CHEMISTRY AND MATERIALS, UNIVERSITY OF BOLOGNA, V. LE RISORGIMENTO 4, 40136 BOLOGNA, ITALY

1 ABSTRACT

Nickel-aluminium mixed hydroxides with different Ni/Al ratios have been prepared by coprecipitation and their homogeneity has been checked by ^{27}Al MAS NMR. An X-ray photoelectron spectroscopic study of calcined precursors before and after leaching with sodium hydroxide solutions has been conducted. The surface of the mixed oxides is found to be richer in aluminium than the bulk composition both before and after NaOH treatment. A model involving the presence of nickel oxide, a Ni-doped alumina phase and a spinel-type phase at the $NiO-Al_2O_3$ interface accounts for the chemical extraction of aluminium and the XPS results.

2 INTRODUCTION

The nickel-aluminium mixed oxides resulting from the thermal decomposition of hydrotalcite-like coprecipitates are among the most studied catalyst precursors for steam reforming and methanation reactions.[1-3] The catalysts obtained by hydrotalcite thermal decomposition followed by reduction exhibit excellent performances owing to their unusually high thermal stability for nickel contents up to 75 Ni wt%.

The thermal decomposition of hydrotalcite-type precursors proceeds in three steps. The lamellar structure of the coprecipitates is observed by XRD in samples calcined in air at temperatures up to 250-300°C[4]. Higher calcination temperatures lead to poorly defined, strongly interacting mixed oxides, sometimes referred to as "NaCl-type mixed metal oxides".[5] The phase separation of pure oxides and spinels finally occurs at high temperatures, typically 700-1000°C. The most interesting calcination range in air is 300-700°C where high specific surface area materials are obtained, and the mixed oxides are very resistant against sintering and reduction. In work described in recent papers, these mixed oxides have been used as supports for metallic catalysts.[6-8] The reasons for this interesting thermal stability are still under debate. In particular, the phases formed as a function of the calcination temperature, and consequently the composition of the mixed oxide surfaces, are not precisely known. The poor cristallinity of these oxides accounts for this uncertainty.

In previous studies, the $NiO-Al_2O_3$ interactions in mixed oxides obtained by calcining hydrotalcite-type coprecipitates have been investigated by employing alkaline washings.[9,10] The existence of a Ni-doped alumina phase in the mixed oxides was likely to be present in view of the loss of more than half of the aluminium initially present in the oxides, NiO being insoluble under these conditions. The structures of the coprecipitates were investigated using thermogravimetric analysis, differential scanning calorimetry and X-ray diffraction. EXAFS was shown to be useful to ensure that the coprecipitation operation did not induce heterogeneities, such as separated aluminium hydroxide or oxyhydroxide phases in the precursors.[10]

In this work, the mixed oxides issuing from the thermal decomposition of nickel-aluminium hydrotalcite-type precursors have been studied by X-ray photoelectron spectroscopy (XPS). The variations of the Ni(2p)/Al(2p) peak intensity ratio as a function of the calcination and washing procedures suggest a surface enrichment in Al in the oxides. The fraction of the surface region attributable to NiO is given by the analysis of the O(1s) level. The XPS analysis strongly supports the previously developed multiphasic model as being responsible for the high thermal stability of the $NiO-Al_2O_3$ mixed oxides. The ^{27}Al MAS NMR is shown to be an effective method for verifying the Ni-Al coprecipitate homogeneity.

3 EXPERIMENTAL

Two nickel-aluminium mixed hydroxides with Ni/Al atomic ratios of 1 (sample A) and 2.8 (sample B) were prepared by coprecipitation at constant pH, 8.0 ± 0.1, by adding a solution of the aluminium and nickel nitrates to a solution containing a slight excess of $NaHCO_3$. The coprecipitates were washed with distilled water then dried in air at 100°C overnight. Only the precursor with Ni/Al = 2.8 (sample B) was subjected to thermal and NaOH treatments since the other precursor was not found to be homogeneous, see text below. Calcined portions of sample B were washed with a 1N sodium hydroxide solution, then rinsed with distilled water to remove nickel hydroxide which had deposited on the washed materials.[10] The NaOH treated samples were finally calcined at the same calcination temperature at which they were heated before NaOH treatment. The Ni and Al contents in the coprecipitates, calcined materials and NaOH-treated samples were determined by X-ray fluorescence spectroscopy, see Table 1. NiO was prepared by calcining Ni(OH)$_2$ at 450°C in air. NiAl$_2$O$_4$ was prepared by calcining a mixture of aluminium and nickel hydroxide in the appropriate stoichiometry at 1100°C in air. A well-crystallized spinel pattern was obtained by XRD and no detectable amount of NiO was observed. NiO and NiAl$_2$O$_4$ were used as reference compounds for the XPS analysis.

^{27}Al NMR spectra were recorded with a Bruker MSL-400 'wide bore' spectrometer at 104.2 MHz using a single pulse sequence. Samples were contained in a 4mm diameter rotor and spun at 11-12 kHz. 1000 FIDs with 4s time intervals were accumulated. XPS experiments were performed on a Kratos Analytical Instrument XSAM-800 spectrometer using MgKα radiation (1253.6 eV).

4 RESULTS AND DISCUSSION

The aim of this work is to show that the thermal decomposition of well defined, homogeneous hydrotalcite-type precursors leads to heterogeneous multiphasic, strongly interacting mixed oxides. Thus, it is essential to ensure that the coprecipitation operation does not induce heterogeneities, such as separated phases of aluminium or nickel (oxy)hydroxide.

<u>Cation Distribution in the Coprecipitates</u>

XRD, TGA and DSC data have been presented in previous works on NiAl coprecipitates with Ni/Al ratios ranging from 0.5 to 3.[9,10] The coprecipitates were found to be homogeneous for Ni/Al atomic ratios ranging from 2 to 3, in accord with other authors[11]. The EXAFS technique provided the confirmation that, for Ni/Al ratios lower than 2, coprecipitation leads to mixtures of a hydrotalcite-type coprecipitate with a Ni/Al ratio equal to 2 and, second, a Ni-free precipitate probably of aluminium (oxy)hydroxide type.[10]

Solid state ^{27}Al NMR is also a useful tool for the verification of the homogeneity of the cation distribution in the brucite type layers. The spectra of coprecipitates A and B are shown in Fig.1. The presence of a separate aluminium hydroxide or oxyhydroxide phase in sample A accounts for a sharp signal 8 ppm downfield from the Al(H$_2$O)$_6^{2+}$ reference (in aqueous Al(NO$_3$)$_3$ solution).[12,13] Conversely, the signal is broadened beyond detection by the paramagnetic Ni^{2+} species when the nickel and aluminium ions are homogeneously distributed in the coprecipitates (sample B, with a Ni/Al ratio equal to 2.8). Thus, ^{27}Al NMR appears to be effective in detecting separate aluminium (oxy)hydroxide patches in NiAl coprecipitates.

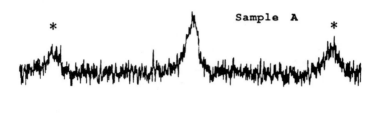
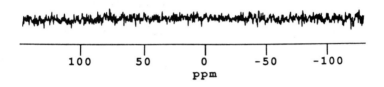

Figure 1 ^{27}Al MAS NMR spectra of samples A and B (chemical shift referenced to an aqueous solution of $Al(H_2O)_6^{2+}$). Spinning side bands are marked by asterisks.

Characterization of Calcined Hydrotalcite-type Precursors

In a previous study where calcined precursors were washed with hot alkaline solutions then rinsed with distilled water, the following results were obtained:
- The amount of Al(III) that passed into solution exhibited a maximum for samples calcined around 700°C: 60% of aluminium initially present in the calcined precursor was dissolved.
- The thermal stability of the NiO particles was not modified by the removal of more than one half of aluminium initially present in the calcined precursors.[10]

Table 1 Physical properties and Ni/Al ratios of the samples as determined by X-ray fluorescence and XPS.

Sample	BET area (m²/g)	Ni/Al bulk atomic ratio*	Ni/Al surface atomic ratio**
A	n.d.	1.0	n.d.
B	n.d.	2.8	2.8
B calc. 450°C	220	2.8	2.2
B calc. 550°C	181	2.8	1.6
B calc. 650°C	153	2.8	1.3
B calc. 450°C and NaOH treated	245	4.0	2.7
B calc. 650°C and NaOH treated	95	6.6	2.7

n.d.: not determined
*: as determined by X-ray fluorescence
**: as determined by XPS

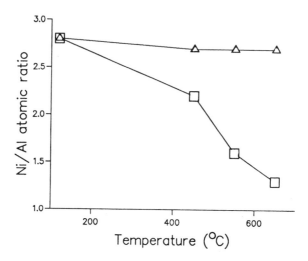

Figure 2 Variation of the $I(Ni_{2p})/I(Al_{2p})$ ratio as a function of the calcination temperature and NaOH treatment. ☐: calcined precursor B before NaOH treatment. Δ: precursor B calcined after NaOH treatment.

The variation of the $I_{Ni\ 2p3/2}/I_{Al\ 2p}$ ratio as a function of the calcination temperature of precursor B is shown in Fig.2. The effect of alkaline leaching is also shown. The measured XPS Ni/Al ratio for the precursor is equal to that found by X-ray fluorescence. One can observe a steady decrease of the Ni/Al ratio with increasing calcination temperature, suggesting a surface enrichment in aluminium. This observation is consistent with a previous observation, indicating that the amount of Al(III) passing into solution increases with the calcination temperature up to 700°C. The surface Ni/Al ratio is boosted after alkaline treatments. This enhancement is obviously due to Al removal. However, the Ni/Al ratio measured by XPS is half of that of bulk material measured by X-ray fluorescence, see Table 1. Thus, the aluminium fraction remaining in the sample is likely to be located on or near the surface of the NiO particles.

The Ni(2p) and O(1s) spectra of NiO and $NiAl_2O_4$ standard compounds and of sample B calcined at 450 and 650°C in air before and after NaOH treatment are shown in Fig.3. The binding energies and widths of the Ni(2p) and O(1s) levels are indicated in Table 2.

The O(1s) level of NiO exhibits a peak at 529.4 eV attributable to oxygen lattice ions and a weaker peak at 531.0 eV assigned to surface OH⁻ groups.[14] Only one O(1s)

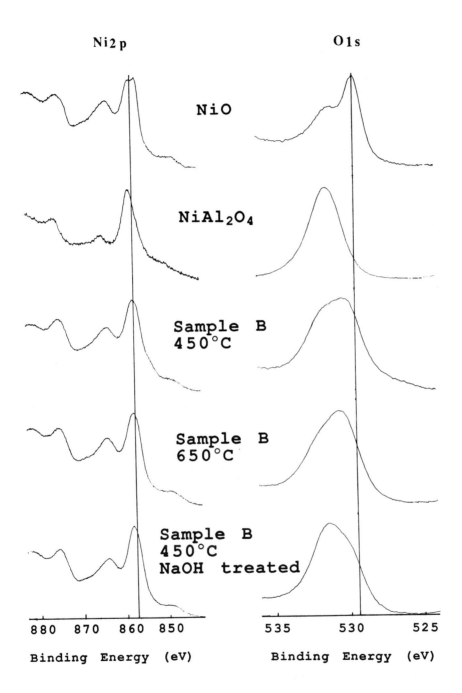

Figure 3 Core level peaks (Ni 2p3/2, O1s) for standard and calcined samples. All binding energies are corrected for charging (C1s reference at 284.6 eV).

peak at 531.3 eV is visible in the spectrum of $NiAl_2O_4$.[15] The O(1s) levels of sample B calcined at 450 and 650°C exhibit a broad peak, which can be resolved in two components at 529.8 and 531.3 eV. Thus, the surface of the calcined hydrotalcite-type precursor appears to be attibutable not solely to NiO, confirming that a substantial fraction of the aluminium is present at the nickel oxide surface. According with the O(1s) spectrum of the $NiAl_2O_4$ standard, the component at 531.3 eV may be attributed to a surface oxygen in a spinel-type phase. However, we have observed similar O(1s) peaks, located from 531.0 to 531.5 eV, for a large number of aluminas in various environments (γ-aluminas, alumina supported NiMo or platinum catalysts, etc.), so that the presence of an alumina phase at the surface of the NiAl mixed oxides may equally well account for the component at 531.3 eV. Assuming that the component at 529.8 eV arises from 'pure' nickel oxide, and that the component at 531.1 or 531.3 eV is due to a spinel-type compound or a Ni doped alumina, the deconvolution of the O(1s) levels into 529.8 and 531.1 peaks gives the fraction of the mixed oxide surface consisting of 'pure' nickel oxide. This fraction, as a function of the calcination temperature and the washing operation, is shown in Figure 4.

About one third of the surface is found to be 'pure' NiO in the calcined materials before NaOH treatment. After NaOH treatment, this fraction is not found to vary substantially. Thus, the aluminium-rich phase present at the surface after alkaline washing is likely to be located at the interface between NiO and the aluminium-rich compound which passed into solution. Even in the NaOH treated materials, i.e. after two thirds of aluminium initially present in the samples has dissolved, the surface of the mixed oxides appears to be mainly of alumina or aluminate type. The surface enrichment in Al is hence confirmed.

Table 2 X.P.S. characteristics of standard compounds and samples

Sample	Ni (2p$_{3/2}$)		O(1s)	
	B.E.(eV)	FWHM*(eV)	B.E.(eV)	FWHM*(eV)
NiO	856.4	4.48	529.4	2.47
$NiAl_2O_4$	858.1	3.86	531.3	2.5
B calc. 450°C	857.7	4.6	529.8	2.5
			531.1	3.2
B calc. 650°C	857.7	4.3	529.8	2.1
			531.3	3.2
B calc. 450°C NaOH treated	858.0	4.2	529.8	2.3
			531.5	2.8
B calc. 650°C NaOH treated	857.8	4.5	529.8	2.1
			531.3	3.1

*: Full width at half maximum peak height

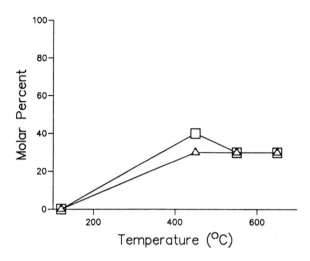

Figure 4 Fraction of the total surface region oxygen attributable to oxygen associated with NiO in calcined precursor B. ☐: calcined precursor B before NaOH treatment. △: precursor B calcined after NaOH treatment.

An examination of the Ni(2p) level permits different chemical forms of nickel in calcined or washed materials to be distinguished. NiO has a doublet structure for the Ni(2p) level and the shake-up satellite is more intense and closer to the main peak than the corresponding satellite of $NiAl_2O_4$. The main Ni(2p) peak location and width as well as the separation and intensity of the shake up satellite in the calcined and NaOH treated samples indicate a mixture of nickel oxide and spinel in these samples (see Fig.3 and Table 2). The binding energies of the $Ni(2p_{3/2})$ levels are closer to that of $NiAl_2O_4$ than to NiO. On the other hand, the shake-up satellite intensities resemble those observed in the NiO spectrum, which clearly indicates the presence of NiO at the surface or next to the surface of the samples. Even though we did not attempt to deconvolute the Ni(2p) levels using the NiO and $NiAl_2O_4$ Ni(2p) levels, the results concerning the Ni(2p) binding energies are consistent with those obtained above. Nickel oxide and a spinel-type compound are both observed in the calcined materials before and after NaOH leaching.

5 CONCLUSIONS

The XPS study is consistent with a previously proposed triphasic model for the calcined NiAl hydrotalcite-type precursors.[10] Examination of the O(1s) and Ni(2p) levels shows that the surface regions of the calcined samples are composed mainly of nickel aluminate or Ni-doped alumina and NiO. The surface Ni/Al ratios as measured by XPS are substantially lower than the bulk Ni/Al ratios, confirming the surface enrichment in Al. The surface Ni/Al ratio is enhanced by the NaOH treatment; however, it remains lower than the bulk value. The Ni(2p) and O(1s) levels are very similar before and after NaOH washing. Thus, the following conclusions can be drawn:
- The surface region in the washed materials is composed of nickel oxide and an aluminium-rich compound, likely of spinel or alumina type.
- NaOH washing removes an aluminium-rich phase, also likely of alumina or aluminate-type. Since the NiO fractions in the washed and unwashed samples are not substantially different, it can be hypothesized that the aluminium-rich compound observed in the NaOH treated materials was located, in the untreated materials, at the interface between NiO and the compound which dissolved.

Previously we suggested that the phase which dissolved was a Ni doped alumina rather than $NiAl_2O_4$, since the amount of Ni(II) which reprecipitated onto the washed samples as nickel hydroxide was negligible in comparison with the amount of Al(III) which passed into solution.[10] Thus, we propose the following model for the mixed oxides obtained by calcination of the NiAl hydrotalcites. The hydrotalcite-type precursors, in which the Ni(II) and Al(III) ions are homogeneously distributed, give upon calcination in air at temperatures from 350 up to 800°C, triphasic mixed oxides consisting of
- a nickel oxide phase, possibly containing Al(III) impurities. The presence of NiO is clearly demonstrated by the XRD,[9,10] EXAFS[10] and XPS data. The particle sizes as measured from XRD measurements range between 35 and 60 Å.
- a spinel-type phase, located at the surface of the NiO particles. The existence of this phase is supported by the XPS data on the NaOH treated materials.
- a Ni-doped alumina phase, probably 'grafted' on the spinel-type phase. This phase can be easily removed by alkaline washing and contributes very slightly to the mixed oxide thermal stability.[10]

We suggest that the spinel type phase is responsible for the thermal properties of the mixed oxides, hindering the growing and sintering of the NiO crystallites when the calcination temperature increases. The spinel type phase also strongly hinders the NiO phase reduction, probably by decreasing the accessibility of the NiO surface to the reducing gas. The spinel type phase may also protect the alumina phase against crystallization, because of its location at the NiO-Al_2O_3 interface. Finally, the existence of aluminum-rich phases, either present before or formed during reduction of the spinel type compound, may explain the properties of the reduced catalysts.

REFERENCES

1. J.R.H. Ross, 'Surface and Defect Properties of Solids', The Chemical Society, Burlington House, London, 1975, Vol.4, Chapter 2, p. 58.
2. J.R.H. Ross, 'Catalysis, Specialist Periodical Reports', Royal Society of Chemistry, London, 1985, Vol.7, p.1.
3. D.C. Puxley, I.J. Kitchener, C. Komodromos and N.D. Parkyns, 'Preparation of Catalysts III', Elsevier, 1983, p.237.
4. F. Cavani, F. Trifiro' and A. Vaccari, Catal. Today, 1991, 11(2), 255.
5. T. Sato, H. Fujita, T. Endo, M. Shimada and A. Tsunashima, React. Solids, 1988, 5, 219.
6. H. Schaper, J.J. Berg-Slot and W.H.J. Stork, Appl. Catal., 1989, 54, 79.
7. R.J. Davis and E.G. Derouane, Nature, 1991, 349, 313.
8. R.J. Davis and E.G. Derouane, J. Catal., 1991, 132, 269.
9. O. Clause, M. Gazzano, F. Trifiro', A. Vaccari and L. Zatorski, Appl. Catal., 1991, 73, 217.
10 O. Clause, B. Rebours, E. Merlen, F. Trifiro' and A. Vaccari, J. Catal., 1992, 133, 231.
11 E.C. Kruissink, L.L. Van Reijen and J.R.H. Ross, J. Chem. Soc., Faraday Trans. 1, 1981, 77, 649.
12 G. Engelhardt and D. Michel, 'High Resolution Solid-State NMR of Silicates and Zeolites', Wiley & Sons, New York, 1987
13 B.H.W.S. De Jong, C.M. Schramm and V.E. Parziale, Geochim. Cosmochim. Acta, 1983, 47, 1223.
14 J. Haber, J. Stoch and L. Ungier, J. Electron Spectrosc. Relat. Phenom., 1976, 9, 459.
15 P. Lorenz, J. Finster, G. Wendt, J.V. Salyn, E.K. Zumadilov and V.I. Nefedov, J. Electron Spectrosc. Relat. Phenom., 1979, 16, 267.

A Supported Copper Catalyst: a Combined XAS and XRD Investigation during Calcination, Reduction, N₂O Decomposition and Catalytic Reaction

G. Sankar[1], J. W. Couves[1], D. Waller[1], J. M. Thomas[1], C. R. A. Catlow[2], and G. N. Greaves[2,1]

[1] DAVY FARADAY RESEARCH LABORATORY, THE ROYAL INSTITUTION OF GREAT BRITAIN, 21 ALBEMARLE STREET, LONDON W1X 4BS, UK
[2] THE SERC DARESBURY LABORATORY, DARESBURY, WARRINGTON, CHESHIRE WA4 4AD, UK

1 INTRODUCTION

A catalyst, prior to reaction, frequently requires activation involving processes such as oxidation, reduction, dehydration or suphidization. Many characterization techniques both *ex-situ* and *in-situ* have been employed to monitor chemical and electronic changes in catalysts during activation processes. Two techniques that give complementary structural information are X-ray absorption spectroscopy (electronic and short-range geometric structures) and X-ray diffraction (long-range periodic structure). Recent developments[1,2] in the instrumentation of both these techniques now allows rapid data acquistion so that *in-situ* measurements of phase changes may be studied in detail in a short interval of time. A typical X-ray absorption (XAS) data collection time is of the order of 50 ms using a photodiode array detector (PDA) and X-ray diffraction (XRD) measurements can be performed within 5 minutes employing a position sensitive detector (PSD).

A fixed geometry experimental arrangement utilizing the PDA[1] and PSD has recently been developed allowing combined XAS and XRD measurements to be made sequentially[2,3]. An environmental cell developed in our laboratory[2] permits us carry out *in-situ* combined XAS/XRD measurements under catalytic reaction conditions as well as allowing the monitoring of the gas products by mass spectrometry or gas chromotography.

The applicability of the combined XAS/XRD/gas analysis technique is illustrated by the following work on the calcination and reduction of a Cu-ZnO based catalyst and the subsequent reverse water-gas shift reaction over the reduced catalyst. Conventional XAS measurements (at room temperature) were also carried out after the activation processes to obtain quantitative information on the nature of the copper particles before and after the dissociative adsorption of N_2O over the reduced catalyst.

2 EXPERIMENTAL

The catalyst precursor was a copper-zinc hydroxycarbonate (possessing the aurichalcite structure) with a Cu/Zn ratio of 20/80, prepared by coprecipitation[4]. The reaction conditions employed during the combined XRD/XAS measurements are described in Fig.1. Gas analysis was carried out using a VG micromass 200 mass spectrometer and a Perkin-Elmer 3500 GC.

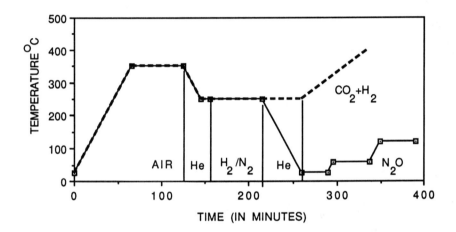

Fig.1 Reaction conditions employed during the course of combined XRD/XAS measurements. The solid line shows the conditions used during dissociative adsorption of N_2O and the dashed line represents the reverse water-gas shift reaction.

The combined XRD/XAS experiments were carried out on station 7.4 at the SERC Synchrotron Radiation Source (SRS), Daresbury laboratory. A schematic diagram of the experimental arrangements on this station is illustrated in Fig.2, the details of which are described elsewhere[2,3].

Room temperature (RT) conventional XAS spectra were taken on station 7.1 of the SRS, Daresbury laboratory. The station was fitted with a Si(111) double crystal monochromator. The same *in-situ* cell as used in the combined experiment was employed and the reaction conditions described in Fig.1 were carried out prior to RT measurements.

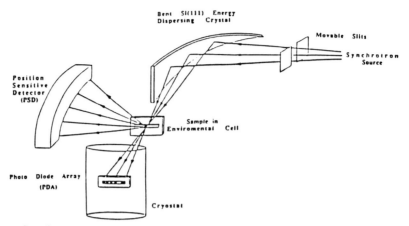

Fig.2 Schematic diagram of the experimental arrangement used for combined XRD/XAS measurement.

The data analysis was carried out using EXBROOK (for pre- and post-edge subtraction) and EXCURV90 (for curve fitting analysis using curved-wave multiple scattering formalism) programs of the Daresbury Laboratory. Cu_2O and Cu metal were used as model compounds for extracting phase-shift information for the curve fitting analysis. The error in estimating coordination number was found to be ±10% and that of the bond lengths ±0.02Å[5].

3 RESULTS AND DISCUSSION

<u>Combined dynamic XAS/XRD studies.</u> In Fig.3(a) we show the XRD patterns recorded during the calcination of the aurichalcite precursor. These demonstrate the formation of CuO and ZnO phases as indicated by the presence of ZnO (100), (002), (101) and (102) reflections and a shoulder on the ZnO (100) peak due to the CuO (111) reflection (Fig.3(b)). However, the CuO (111) reflection is not as distinct as in the earlier study[2], since the calcination temperature was restricted to 350°C in the present work as opposed to 450°C. It is interesting to note that during the calcination of aurichalcite, the (610), (10,00) and (620) reflections disappear rapidly whilst the (511) and (711/420) reflections diminish slowly. This demonstration of the anisotropic decomposition of aurichalcite shows the advantage that the short measurement time scale of this technique has over conventional XRD measurement. The XAS data did not show any appreciable change during the calcination, as with our earlier observation[2] since the oxidation state of copper is expected to be the same prior to and after calcination. It is difficult to obtain quantitative structural information from the EXAFS data

A Supported Copper Catalyst: a Combined XAS and XRD Investigation

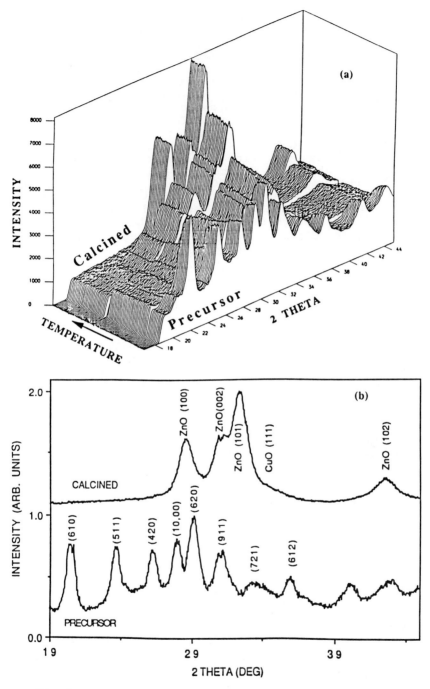

Fig.3 (a) XRD patterns recorded during calcination of the aurichalcite precursor; (b) Indexed XRD plots of the precursor and the calcined catalyst

because of the restrictions dictated by the temperature of the measurement as well as the low resolution.

The XRD/XAS data recorded during the reduction are shown in Fig.4. The XRD pattern (Fig.4(a)) shows that the reduction leads to the disappearance of CuO(111) shoulder and the emergence of the Cu(111) peak of Cu metal; the reflections corresponding to the ZnO phase were found to be unaffected. The apparent decrease in the ZnO(101) reflection is due to the disappearance of the overlapping CuO(110) reflection. The particle size estimated from the line broadening of the Cu(111) reflection, using the Scherrer equation, was found to be in the region of 75 to 100Å. The ED EXAFS data (Fig.4b) clearly support the reduction of CuO to Cu metal, as inferred from the edge-shift as well as the change in the white-line intensity. The ED EXAFS did not show features characteristic of Cu^{1+} suggesting a direct reduction from Cu^{2+} to metallic copper.

The activity of the catalyst towards the reverse water-gas shift reaction (CO_2:1, H_2:1, N_2:8, at a flow rate of 50ml/min) was determined whilst raising the temperature of the reduced catalyst from 260°C to 400°C. The CO_2 turnover numbers are shown in Fig.5. The turnover numbers were found to be lower than the micro-reactor studies due to the geometry of the environmental cell which gives rise to dilution of the reaction product by the incoming reactants. However, such reaction measurements are still worthwhile as they do indicate that the diffraction and spectroscopic measurements are of an active catalyst.

Conventional high resolution XAS studies. In view of the XRD/XAS results, we performed high resolution Cu K-edge XAS measurements in order to obtain quantitative structural information as to the nature of the copper particles in the reduced state and after dissociative adsorption of N_2O. The Fourier transform (FT) of the Cu K-edge EXAFS of the reduced catalyst shows (Fig.6a) features similar to bulk copper in the FCC structure. Analysis of the EXAFS data shows that the Cu-Cu distances were contracted by about 0.006Å accompanied by reduction in the coordination number. The structural data obtained from the analysis are listed in Table 1. The Cu particle size was estimated by comparing the coordination number obtained from the experimental data with that calculated by using an algorithm similar to that reported by Greegor and Lytle[6]. The results indicate a particle diameter in this catalyst of ca 25Å. The small contractions in the Cu-Cu distances are also consistent with the small particle size.

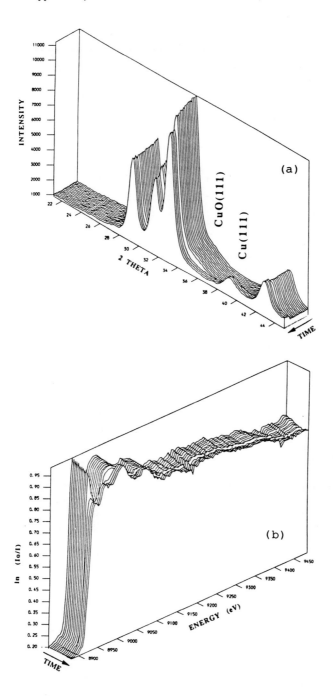

Fig.4 Combined (a) XRD and (b) Cu K-edge XAS recorded sequentially during the reduction of the calcined catalyst.

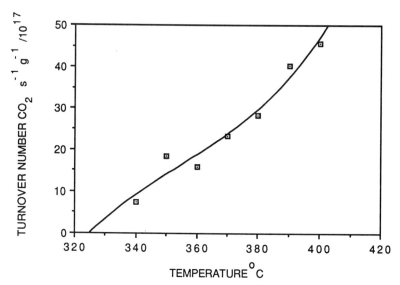

Fig.5. CO_2 turnover number plotted against temperature. Note that the data were obtained from mass spectrometer and GC collected during an *in-situ* reaction while recording the combined XRD/XAS data.

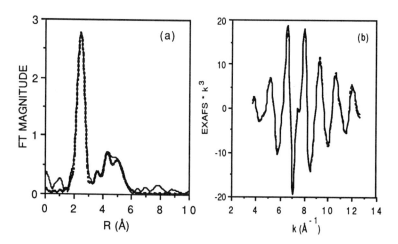

Fig.6 (a). FT of the Cu K-edge EXAFS of the reduced catalyst. (b) Fourier filtered EXAFS (in the range 0.7 to 5.8Å) of the reduced catalytst. The solid line denotes the experimental data and the dashed line represents the calculated EXAFS using parameters listed in Table 1.

The FT of the Cu K-edge EXAFS (Fig.7(a)) recorded after performing dissociative adsorption of N_2O, at

Table 1: Structural parameters obtained from the analysis of our EXAFS data (recorded by conventional means).

	Atom-Pair	N	R(Å)	$\Delta\sigma^2$ (a)
Reduced	Cu-Cu	9.9	2.55	0.0006
	Cu-Cu	3.9	3.61	0.0010
	Cu-Cu	15.1	4.42	0.0008
	Cu-Cu	7.5	5.11	0.0009
	Cu-Cu	12.7	5.71	0.0010
N$_2$O treated	Cu-O	0.4	1.89	0.0020
	Cu-Cu	8.2	2.55	0.0006
	Cu-Cu	3.9	3.61	0.0010
	Cu-Cu	11.7	4.42	0.0008
	Cu-Cu	5.9	5.11	0.0009
	Cu-Cu	12.4	5.71	0.0010

(a) $\Delta\sigma^2$ refers to the difference between the σ^2 of the catalyst and the model compound (Cu metal in the case of Cu-Cu atom-pair and Cu$_2$O for Cu-O atom-pair).

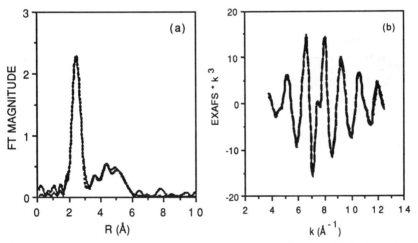

Fig.7 (a) FT of the Cu K-edge EXAFS of the N$_2$O treated sample. (b) Fourier filtered EXAFS (in the range 0.7 to 5.8Å) of the N$_2$O treated catalyst. The solid line shows the experimental data and the dashed line represents the calculated EXAFS using parameters listed in Table 1.

60°C, over the reduced catalyst shows a dimunition in the intensity of the peaks related to Cu-Cu distances. However, we do not see the expected peaks associated with Cu-O scattering from the oxidized phase. The analysis of the EXAFS data shows that there is a decrease in the Cu-

Cu coordination numbers (by approximately 18 percent); furthermore the statistical analysis performed after incorporating a Cu-O distance of ca1.89Å (significant at a level <5%)[7] confirms the existence of a surface oxide phase. A separate experiment[8] carried out on highly dispersed Cu/SiO$_2$ (particle size of ca13Å) showed clearly the presence of a Cu-O distance of 1.87Å which was accompanied by a decrease in Cu metal coordination (by 65 percent) supporting the present EXAFS results. The decrease in the Cu metal coordination number arises from the presence of two different coordination shells similar to that observed on oxidized polycrystalline copper[9]. Based on the Cu-O distance obtained from the analysis we[8] were able to propose a long bridged Cu-O-Cu on a Cu(110) surface as a model for the oxygen adsorption site after dissociative adsorption of N$_2$O. It appears that a oxygen is occupying a similar site on the surface of the copper particles studied here.

4 ACKNOWLEDGEMENTS

The authors wish to thank Drs A.J.Dent and G.E. Derbyshire for technical help with the combined experiments. We are grateful to SERC for provision of beam time and support of this project. One of the authors (GS) wishes to thank BP for the financial support. We would like to acknowledge the use of SERC funded Chemical Data Service at Daresbury Laboratory for part of the work described in this paper

5 REFERENCES

1. N.M. Allison, G. Baker, G.N. Greaves and J.K. Nicoll, Nucl. Instrum. Meth., 1988, A266, 592.
2. J.W. Couves, J.M. Thomas, D. Waller, R.H. Jones, A.J. Dent, G.E. Derbyshire and G.N. Greaves, Nature, 1991, 354, 465
3. J.M. Thomas, J. Chem. Soc. Dalton Trans., 1991, 555
4 D. Waller, D. Stirling, F.S. Stone and M.S. Spencer, Farad. Disc. Chem. Soc., 1989, 87, 107
5. S.J. Gurman, 'Applications of Synchrotron Radiation', eds. C.R.A. Catlow and G.N. Greaves, Blackie, Glasgow, 1990
6. R.B. Greegor and F.W. Lytle, J. Catal., 1980, 63, 476
7. R.W. Joyner, K.J. Martin and P. Meehan, J. Phys. C, 1987, 20, 4005.
8. G. Sankar, J.M. Thomas, D. Waller, J.W. Couves, C.R.A. Catlow and G.N. Greaves (submitted for publication)
9. S. Pizzini, K.J. Roberts, G.N. Greaves, N.T. Barrett, I. Dring and R.J. Oldman, Farad. Disc. Chem. Soc., 1990, 89, 51.

Lability of Small Rhodium Particles in Carbon Monoxide: Influences of Chlorine and Particle Size

P. Johnston and R. W. Joyner
LEVERHULME CENTRE FOR INNOVATIVE CATALYSIS, DEPARTMENT OF CHEMISTRY, UNIVERSITY OF LIVERPOOL, LIVERPOOL L69 3BX, UK

1 INTRODUCTION

The lability of small rhodium particles in carbon monoxide was first demonstrated conclusively in an EXAFS study by Koningsberger, Prins and colleagues[1], (hereafter referred to as KP), and has been confirmed by others[2-6]. Exposure to CO at room temperature resulted in almost complete disruption of small rhodium particles, diameter < 10A, and the formation of species identified as Rh(I) *gem*-dicarbonyl. We were originally interested in the stability of this species in synthesis gas, a mixture of CO and hydrogen, and therefore undertook an *in situ* EXAFS study that has been reported before[6]. We deliberately prepared catalysts which were as similar as possible to those made by KP and indeed observed the same disruption of the rhodium particles. After exposure to CO the nearest neighbour Rh - Rh coordination number fell from 5.5 to 0.3. The conclusions of our analysis of similar EXAFS results differed from that of KP in that we assigned an interatomic distance of ca 2.3A to Rh - Cl bonding, rather than Rh-O. Our single scattering analysis was criticised[7], as multiple scattering may be significant for carbonyl compounds, particularly for the Rh - O distance.

In all of the studies where disruption of rhodium particles has been demonstrated by EXAFS, the catalyst has been prepared from a chloride precursor. In view of this and the possibility of chlorine involvement in the catalyst particle reactivity[8], we have carried out additional experiments designed to clarify the possible role of chlorine, which is agreed to be retained by the catalyst even after reduction[1,6], and to study further the stability of small rhodium particles in carbon monoxide. Catalysts exhibiting similar small rhodium particles have been made from chloride and nitrate precursors and their stability in carbon monoxide examined; catalysts with different particle size made from a chloride precursor have also been studied.

2 EXPERIMENTAL

Catalyst preparation and materials for the chloride containing catalysts have been described previously[6]. A catalyst was also prepared from a solution of Rh(NO)3 in 10% HNO3, supplied by Johnson Matthey PLC. This was introduced to the same γ-alumina as used for the chloride containing catalysts, using an incipient wetness method. Three catalysts were studied, two of which were made using rhodium chloride as the precursor and the other using rhodium nitrate. EXAFS measurements were performed on line 9.2 at the Daresbury synchrotron radiation source, using an *in situ* cell that has been described before[9]: samples were studied in the form of 13 mm diameter pressed discs. Before exposure to carbon monoxide, samples were reduced in flowing hydrogen (50 ml min^{-1}) at 493 K, cooled to room temperature and examined by X-ray absorption. The hydrogen flow was halted and the gas pumped away; the catalysts were then exposed to static carbon monoxide at 1 bar pressure. The sample holder allowed three catalysts to be treated at the same time and then studied sequentially by X-ray absorption. Techniques of data analysis have been described before[6]. The significance of shells in the EXAFS analysis has been validated by statistical techniques[10] using unfiltered data. The spectra presented have been subjected to Fourier filtering to improve the quality of background subtractions. Phase shifts were obtained from the Daresbury data base and optimised by analysis of spectra from rhodium foil and $(Rh(CO)_2Cl)_2$, in the range 3 - 16 Å^{-1}.

Catalytic data were measured on a laboratory microreactor equipped with mass flow controllers and with GC analysis of products. The catalyst charge was 0.2 g and the reactor internal diameter was 6 mm.

3 RESULTS

The rhodium K shell EXAFS for the reduced catalysts after background subtraction and Fourier filtering are shown in Figure 1 and the results of the data analysis are listed in Table 1. Two catalysts made from chloride precursors were studied (Figs. 1a,c), and one made with rhodium nitrate (Fig. 1b). For the first chloride precursor catalyst the surface chlorine / rhodium ratio was determined by X-ray photoelectron spectroscopy to be 1.7, indicating substantial retention of chloride by the catalyst, in agreement with previous observations[1,6].

The results of exposing the catalysts to carbon monoxide at room temperature and 1 bar pressure are shown in Fig. 2, and the results of the data analysis are listed in Table 2.

Catalytic measurements were carried out at 523 K in a 1:1 ratio of CO/H_2, as has been described previously[6]. The product distribution has also been described before and is typical of rhodium / alumina catalysts,

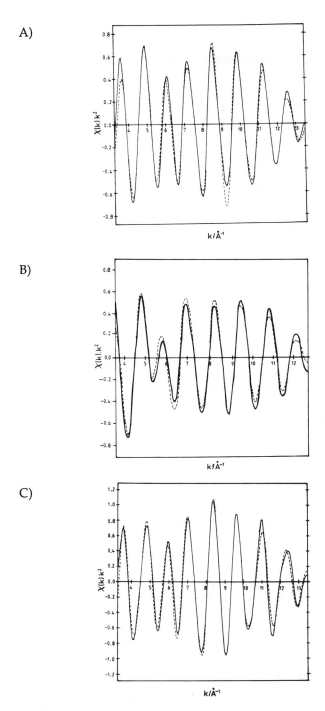

Figure 1 Rh K shell EXAFS from catalysts reduced at 493 K; solid lines, experimental; dashed lines, calculated using the parameters given in Table 1. A) Chloride precursor, catalyst No. 1; B) Nitrate precursor; C) Chloride precursor, catalyst No. 2.

Table 1 EXAFS Results for the Reduced Catalysts

A) Chloride Precursor, Catalyst Number 1

Neighbour	Interatomic Distance / Å	Coordination Number	Debye-Waller Factor/ Å^2
Rhodium	2.64 ± 0.02	4.6 ± 0.5	0.015
"	3.70	1.3	0.019
"	4.56	1.9	0.019
Chlorine	2.34	0.7	0.013

B) Nitrate Precursor

Neighbour	Interatomic Distance / Å	Coordination Number	Debye-Waller Factor/ Å^2
Rhodium	2.68 ± 0.02	4.3 ± 0.5	0.015
"	3.69	1.1	0.018
"	4.62	1.7	0.015

C) Chloride Precursor, Catalyst Number 2

Neighbour	Interatomic Distance / Å	Coordination Number	Debye-Waller Factor/ Å^2
Rhodium	2.69 ± 0.02	6.5 ± 0.5	0.015
"	3.75	2.9	0.013
"	4.66	2.0	0.020

containing a mix of hydrocarbons and oxygenates. The performance of chloride-containing and chloride-free catalysts has been compared over an extended period. These results will be described in detail elsewhere, but one evident trend is illustrated in Figure 3. For the chloride-free catalyst selectivity to hydrocarbons is approximately constant with time on stream: for the chloride-containing catalyst hydrocarbon selectivity declines significantly during the first day of catalyst operation.

4 DISCUSSION

Rhodium particles formed from the reduction of the chloride and nitrate precursor materials are very similar in size, as can be seen from comparison of the spectra shown in Fig. 1a and b, and the results in Table 1. For the chloride precursor catalyst No. 1 the nearest neighbour Rh-Rh coordination number is 4.6 ± 0.5, corresponding to particles with an average

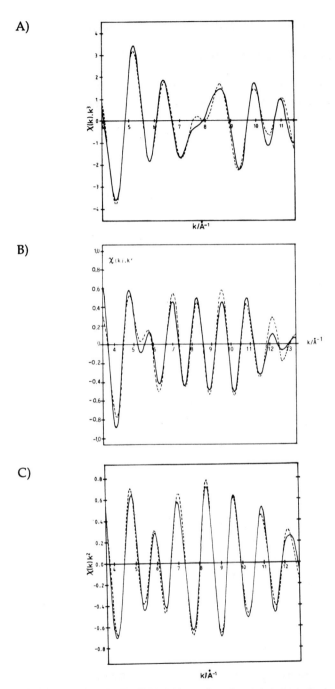

Figure 2 Rh K shell EXAFS from catalysts after reduction in hydrogen at 493 K and exposure to carbon monoxide at 298 K; solid lines, experimental; dashed lines, calculated using the parameters given in Table 1. A) Chloride precursor, catalyst No. 1; B) Nitrate precursor;
C) Chloride precursor, catalyst No. 2.

Table 2 EXAFS Results for the Catalysts in Carbon Monoxide

A) Chloride Precursor, Catalyst Number 1

Neighbour	Interatomic Distance / A	Coordination Number	Debye-Waller Factor/ A^2
Rhodium	2.68 ± 0.04	0.7 ± 0.3	0.015
Carbon	1.83 ± 0.05	1.4 ± 0.4	0.008
Chlorine	2.20	0.7	0.009
Oxygen	2.68	0.7	0.005

B) NitratePrecursor

Neighbour	Interatomic Distance / A	Coordination Number	Debye-Waller Factor/ A^2
Rhodium	2.70 ± 0.02	3.9 ± 0.5	0.015
"	3.69	0.5	0.013
Carbon	1.96	2.2	0.012

C) Chloride Precursor, Catalyst Number 2

Neighbour	Interatomic Distance / A	Coordination Number	Debye-Waller Factor/ A^2
Rhodium	2.69 ± 0.02	5.4 ± 0.5	0.016
"	3.68	2.9	0.016
"	4.66	2.0	0.020
Carbon	0.83	2.02	0.014

of 10 rhodium atoms, while that for the nitrate precursor is 4.3 ± 0.5, indicating an average particle size of 9 atoms. For the chloride precursor material there is some evidence in the EXAFS of Rh - Cl bonding, with a coordination number of 0.7 ± 0.3 and a characteristic interatomic distance. No distances other than Rh-Rh are statistically significant for the catalyst prepared from the nitrate. As noted previously[11], particles adopt the bulk, face centred cubic structure, as evidenced by the presence of the appropriate non-nearest neighbour coordination distances.

Marked differences are observed when the two catalysts are exposed to CO. As noted previously, nearly complete disruption of the particles of the chloride precursor No. 1 catalyst is observed. The Rh - Rh nearest neighbour CN falls to 0.7 ± 0.3, while coordination to CO is clearly evident. Analysis of the spectrum *with the inclusion of multiple scattering* indicates the following: each rhodium is bound, on average, to 1.4 molecules of CO and

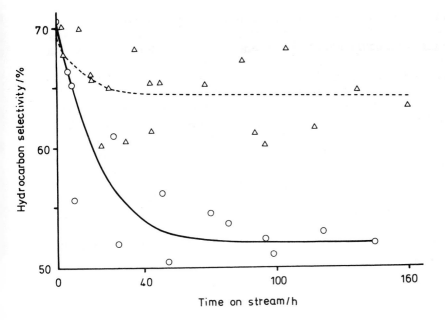

Figure 3 The selectivity to hydrocarbons from synthesis gas of rhodium / alumina catalysts prepared from diferent catalysts; test conditions are given in the text. Circles and the solid line show the results from two catalysts prepared from chloride precursors; the selectivity after 100 h under test is 52 ± 2%. The triangles and the dashed lines show the results from three runs on catalysts prepared from the nitrate precursor; the selectivity after 100 h under test is 64 ± 3%. The selectivity of the chloride precursor materials is given in more detail in Reference 6.

0.7 atoms of chlorine. As we predicted[12], inclusion of multiple scattering in the EXAFS data analysis does not necessitate any modification of our previous conclusions that the central peak in the Fourier transform of the EXAFS spectrum is due to Rh - Cl bonding and is not adequately described by a Rh - O bond.

The change on exposure to CO of the catalyst made from the nitrate precursor is much less marked. The Rh - Rh nearest neighbour coordination number falls from 4.3 ± 0.5 to 3.7 ± 0.5, a change that is within experimental error but could reflect a decrease in the number of atoms in the average particle from 9 to ca 7.5. The EXAFS spectrum shows that the particles have interacted strongly with the carbon monoxide, as can be seen from the emergence of a Rh - C distance with a coordination number of ca 2. The resultant composition is much more like that of a rhodium carbonyl, for example $Rh_6(CO)_{16}$.

The chlorine containing catalyst is thus readily disrupted by carbon monoxide while that where chlorine is absent is not. As we originally argued, it seems that chlorine has a role in the stabilisation of the *gem*-dicarbonyl species, as it does in the crystalline compound di-rhodium tetra-gem-dicarbonyl dichloride. In the absence of chlorine the metal particles adsorb substantial amounts of carbon monoxide, but are much less substantially disrupted.

The results from catalyst chloride precursor No. 2 indicate that the lability of rhodium particles also depends on size. The reduced particles are about twice as large as those formed in chloride precursor No. 1, with a Rh-Rh nearest neighbour coordination number of 6.5 ± 0.6 indicating the presence of ca 18 atoms per cluster. These particles are disrupted by CO, with the average size falling to 13 atoms (Rh-Rh CN of 5.4 ± 0.5), but the effect is much less marked than for chloride precursor No. 1. Again there is evidence for chemisorbed CO, with a Rh-C coordination number of ca 1.1. We believe that this catalyst is not monodisperse and that the result reflects the disruption predominantly of the smaller particles while the larger ones remain intact.

Figure 2 shows that the residual chlorine present on the catalyst has some influence on the catalytic performance. Where no chlorine is present the selectivity to hydrocarbons remains approximately constant with time on stream, at $64 \pm 3\%$. When catalysts are made from chloride precursors, the hydrocarbon selectivity decreases during the first 20 - 40 hours on stream, reaching a constant value of $52 \pm 2\%$. The reasons for these changes are not clear, but we hope to probe them in more extensive future *in situ* EXAFS studies.

5 ACKNOWLEDGEMENTS

We are grateful for the support of this work by the Science and Engineering Research Council and by the Industrial Affiliates of the Leverhulme Centre. We are grateful to Prof. E.S. Shpiro for experimental assistance.

6 REFERENCES

1. H.F.T. Van't Blik, J.B.A.D. Van Zon, T. Huizinga, J.C. Vis, D.C. Koningsberger and R. Prins, J. Phys. Chem., 1983, 87, 2264; J. Mol. Cat., 1984, 25, 379.
2. B.G. Frederick, G. Apai and T.N. Rhodin, J. Amer. Chem. Soc., 1987, 109, 4797.
3. K. Tamaru and Y. Nihei, J. Catal., 1989, 115, 273.
4. A. Erdohelyi and F. Solymosi, J. Catal., 1983, 84, 446; ibid, 1988, 110, 413.
5. P. Basu, D. Panyatov and J.T. Yates, J. Phys. Chem., 1987, 91, 3133.
6. P. Johnston, R.W. Joyner, P.D.A. Pudney, E.S. Shpiro and B.P. Williams, Faraday Disc. Chem. Soc., 1990, 89, 91.
7. D.C. Koningsberger, ibid, 1990, 89, 144.
8. A. Crucq, L. Degols, A. Frennet and E. Lienard, J. Mol. Cat., 1990, 59, 257.
9. R.W. Joyner and P. Meehan, Vacuum, 1983, 33, 691.
10. R.W. Joyner, K.J. Martin and P. Meehan, J. Phys. C, Colid State Phys., 1987, 20, 4005.
11. P. Johnston, R.W. Joyner and P.D.A. Pudney, J. Phys. Condensed Matter, 1989, 1, SB171.
12. R.W. Joyner, Faraday Disc. Chem. Soc., 1990, 89, 148-9.

The Advantageous Use of Microwave Energy in Catalyst Characterisation

G. Bond, R. B. Moyes, and D. A. Whan
SCHOOL OF CHEMISTRY, UNIVERSITY OF HULL, HULL HU6 7RX, UK

1 INTRODUCTION

The origins of temperature programmed techniques date back to the work of Amenomiya and Cvetanovic[1], who developed temperature programmed desorption, and that of Robertson et al.[2] who carried out the early work on temperature programmed reduction (TPR). Other similarly related methods include temperature programmed oxidation (TPO) and temperature programmed decomposition. These techniques are related in that they can be carried out using the same equipment, comprising essentially a thermal conductivity cell and associated electronics, a linear temperature programmer, a sample holder, a furnace and cold traps. These techniques are employed extensively in the characterisation of catalysts and their popularity lies in the wealth of information that can be obtained from relatively inexpensive equipment.

The main drawback with conventional temperature programmed techniques is that they rely on efficient heat transfer from the wall of the furnace to the sample holder and then from the sample holder to the sample. Heat transfer in this type of system is seldom achieved with a high degree of efficiency, hence temperature gradients are established within the catalyst bed thus inevitably resulting in output of poor resolution which can lead to the loss of valuable information.

Microwave energy provides a means of heating catalyst samples more homogeneously than by conventional means. The use of a continuously variable power microwave source allows the temperature to be raised linearly and more uniformly than has been previously possible. Temperature programmed techniques using microwave energy to provide the heating provide a means of increasing the resolution of these well-established methods.

2 EXPERIMENTAL

When assessing the effect of microwave radiation on any process, it is essential to draw a direct comparison with the same process using a conventional furnace or oven. All other equipment and its geometry should be identical for both experiments. In this study, conventionally heated samples have been heated using a furnace and temperature programmer (Stanton Redcroft type 706) which is specially designed for temperature programmed studies. The apparatus used to heat the sample using microwave radiation was a single mode cavity constructed to our specification by Microwave Heating Ltd. The microwave frequency employed was 2.45 GHz. Apart from the source of energy for heating, all other equipment was identical.

Design and Temperature Control within the Single Mode Cavity

Most people are now familiar with the domestic microwave oven which operates as a multimode cavity. Microwaves are generated by a magnetron and pass to the cavity (food compartment) by way of a short section of wave-guide. The cavity may contain a mode stirrer to aid reflection of the microwaves in a random manner, with the aim of producing a uniform microwave field within the cavity. Devices of this type have a crude means of power control which consists of switching the microwave supply on and off. The time scale of this switching is, however, of the order of several seconds, thus making this form of equipment unsuitable for reactions in flowing systems.

The single mode cavity used in the present study (Figure 1) can be considered to consist of five main units. These are a control unit which is capable of continuously varying the microwave power, a microwave generator (magnetron, frequency 2.45 GHz, maximum power 750 W), a section of wave-guide into which the reactor is placed, a means of monitoring the temperature while the sample is being irradiated, and a personal computer with suitable analogue to digital capabilities.

The wave-guide can itself be considered as five separate sections (Figure 2). Microwaves produced by the magnetron enter the launch section and then the test section through the circulator. The function of the circulator is to reflect the microwaves through 120°. This causes microwaves exiting the launch section to be directed into the test section, and microwaves which are not absorbed by the sample to be turned through a further 120° into a water load. This prevents reflected microwaves re-entering the launch section where they would cause overheating and eventual failure of the magnetron. The function of the contactless double plunge tuner is to reflect the microwaves efficiently, thus establishing a standing wave with the maximum in E field at the point at which the sample is mounted in the test section. Tuning of the cavity is achieved by varying the position of the double plunge tuner, by way of a screw

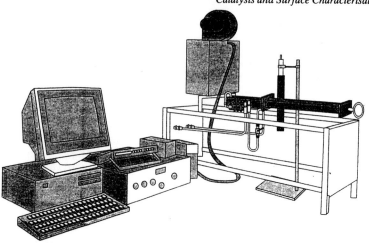

Figure 1. Diagrammatic representation of single mode cavity and associated control equipment.

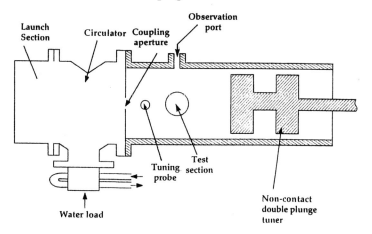

Figure 2. Plan view of wave-guide section of apparatus.

thread mechanism, so that a maximum signal is obtained from a tuning probe.

The sample under investigation was mounted in the test section. The temperature of the sample may be monitored either by a gas thermometer, built into the vessel, or by viewing the sample with an infrared pyrometer through the observation port. The use of gas thermometry for the measurement of temperature within a microwave field has been dealt with in more detail elsewhere[3,4]. The experimental results presented here have all made use of an infrared pyrometer (Tasco THI-302DX) to measure the sample temperature. The pyrometer responds to infrared radiation over the range 8 - 12 μm and is capable of measuring temperature over the range 0 - 500°C. Over this temperature range, the pyrometer yields an output of 0 - 5 V which is digitised and fed to the computer. At frequent intervals the computer calculates the desired temperature, according to the time since the start of the experiment and the required ramp rate. The difference between the measured temperature value and the computed value is then used to calculate the required analogue voltage which the computer sends to the control unit, thus varying the microwave power. The computer reads the temperature and adjusts the power of the microwave radiation on average once per second. Control at not less than this frequency is necessary as the dielectric properties of materials are usually temperature dependent and hence their heating characteristics continually change as samples are heated.

Sample Preparation

A copper/zinc/aluminium hydroxycarbonate material with a Cu:Zn:Al ratio of 6:3:1 was prepared as previously described by Höppener et al[5]. from solutions of copper(II) nitrate (1 M), zinc nitrate (1 M) and aluminium nitrate (1 M) and sodium carbonate (2 M). The precipitation was carried out at 80°C and a constant pH of 7. The precipitate was filtered and washed with distilled water before drying. The phases present in the dried material were determined using X-ray powder diffraction.

Temperature Programmed Decomposition

Temperature programmed decompositions were carried out at heating rates of 5°C min^{-1} and 15°C min^{-1}. The reaction vessel was constructed with flat faces of diameter 15 mm, the distance between the faces being 3 mm. The reason for this design was to provide a sufficiently large area on which to focus the infrared pyrometer. The mass of sample required to fill the reaction vessel was 0.13 g.

The experimental configuration was such that the helium carrier gas (25 ml min^{-1}) travelled through a cold trap before passing through the reference side of the thermal conductivity detector (TCD). The carrier gas was then directed through the reactor which was mounted either in the single mode cavity (microwave heating) or the tube furnace (conventional heating). The exit gas from the reactor then passed through a further cold trap before entering the second arm of the TCD. The signal from the TCD and the temperature reading were stored by the computer. The computer also stored the microwave power level: this can be a useful parameter to record as phase changes in the sample frequently result in changes in the dielectric properties of the material under examination. This is seen as a sudden change in the microwave power required to maintain the required rate of temperature increase.

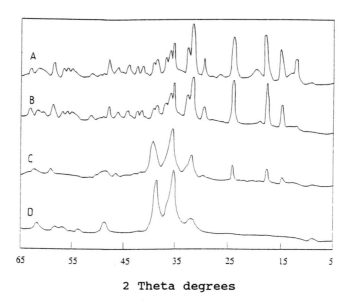

Figure 3. Powder X-ray diffraction patterns of starting material and products at various stages of decomposition (see text for details).

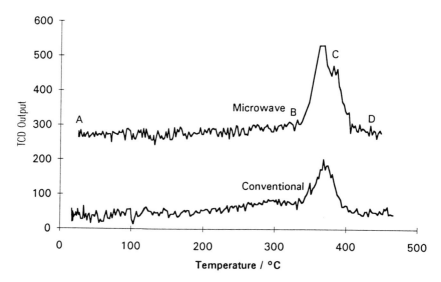

Figure 4. Temperature programmed decompositions of a Cu/Zn/Al hydroxycarbonate using a heating rate of 15°C per minute.

3 RESULTS

Examination of the dried precipitate by X-ray powder diffraction (Figure 3A) reveals that the main phases are primarily malachite ($CuCO_3 \cdot Cu(OH)_2$) and a lesser amount of the copper zinc aluminium carbonate hydroxide hydrate ($Cu_2Zn_4Al_2(OH)_{16}CO_3 \cdot 4H_2O$) which possesses the hydrotalcite structure.

Figure 4 shows the decomposition profiles for the 15°C min^{-1} heating rates. The decomposition profile obtained using microwave radiation clearly shows more structure than the comparative experiment performed using a conventional furnace.

The increased structure prompted us to investigate the phases which were responsible for each of the peaks in the thermal decomposition profile. Separate microwave experiments were performed, again using the 15°C heating rate, and were terminated at temperatures corresponding to the points B, C and D shown on Figure 4. As previously, the samples were analysed using X-ray diffraction to determine the phase changes which had taken place. The diffractograms B, C and D in Figure 3 correspond respectively to heating to points B, C, and D in Figure 4

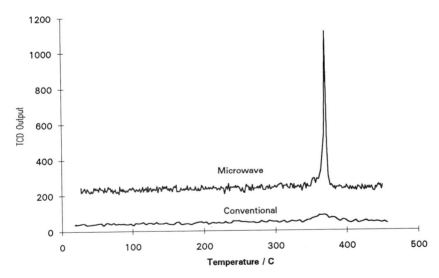

<u>Figure 5</u>. Temperature programmed decompositions of a Cu/Zn/Al hydroxycarbonate using a heating rate of 5°C per minute.

Comparing the X-ray diffraction pattern obtained from heating to point B with that of the starting material, the main change is the loss of the peak situated at a 2 theta value of 11.75° in diffractogram A. This peak is due to the 003 reflection from the hydrotalcite material. It therefore appears that the early stages of the decomposition profile correspond to the disappearance of the hydrotalcite phase. Heating the material to points C and D shows loss of the malachite phase and the formation of the copper and zinc oxides. However, the X-ray diffractograms do not yield any useful information as to why the main peak in the decomposition profile should be made up of two narrower, overlapping peaks.

Figure 5 shows the decomposition profiles for the experiments performed using a heating rate of 5°C min^{-1}. The decomposition obtained using a conventional furnace resulted in a very broad peak and consequently the height of the peak was extremely low. The same experiment performed using microwave radiation gave a much sharper peak than that observed in the conventionally heated experiment.

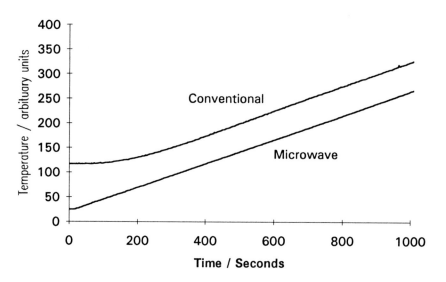

Figure 6. A comparison of increase in temperature with time for microwave and conventional heating

4 DISCUSSION

It is clear that the more uniform heating obtainable with microwave radiation, using the apparatus described within this paper, yields a significant improvement on a typical temperature programmed technique using conventional heating.

One of the advantages of microwave heating, over conventional heating, which is not shown up in the results is the more linear heating that is obtained by the use of microwaves. Temperature programmed heating using a conventional furnace always results in a temperature lag especially in the early stages of heating, as illustrated in Figure 6. The temperature profile from the microwave experiment clearly indicates the more linear heating obtainable within the microwave system. The microwave system also has the added advantage that there is no need to wait for a furnace to cool before the next sample is run. Therefore the throughput of samples can be much higher in the microwave system.

The more uniform heating which is possible using microwave radiation results in sharper peaks that can increase the amount of information available from temperature programmed studies. These sharper peaks result in a marked increase in sensitivity when microwave heating is employed.

The use of temperature programmed studies using microwave radiation for the study of catalysts appears to offer many advantages over conventional heating.

REFERENCES

1. Y. Amenomiya and R.J. Cvetanovic, J. Phys. Chem., 1963, 67, 144.
2. S.D. Robertson, B.D. McNicol, J.H. de Bass and S.C. Kloet, J. Catalysis, 1975, 37, 424.
3. E. Karmazsin, R. Barhoumi and P. Satre, J. Thermal Anal., 1984, 29, 1269.
4. G. Bond, R.B. Moyes, S.D. Pollington and D.A. Whan, Measurement Sci. and Technol., 1991, 2, 571.
5. R.M. Höppener, E.B.M. Doesburg and J.J.F. Scholten, Applied Catalysis, 1986, 25, 109.

Inelastic Neutron Scattering Studies of Catalysts – Thiophene Adsorbed on Sulphided Molybdenum/Alumina

Philip C. H. Mitchell[1], Jean Grimblot[2], Edmond Payen[2], and John Tomkinson[3]

[1] DEPARTMENT OF CHEMISTRY, THE UNIVERSITY, WHITEKNIGHTS, READING RG6 2AD, UK
[2] LABORATOIRE DE CATALYSE, UNIVERSITÉ DES SCIENCES ET TECHNIQUES DE LILLE FLANDRE-ARTOIS, 59655 VILLENEUVE D'ASCQ CEDEX, FRANCE
[3] RUTHERFORD APPLETON LABORATORY, CHILTON, DIDCOT, OXFORDSHIRE, OX11 0QX, UK

1 INTRODUCTION

Catalysts based on molybdenum disulphide are used in hydrotreating petroleum crudes, catalysing the reaction of organo-S, -N, and -O compounds with hydrogen and, thereby, their removal from the crude.[1] Typically the catalyst, commonly referred to as a hydrodesulphurisation catalyst, consists of 8% Mo and 3% Co, as sulphides, supported on γ-alumina. The active phase comprises molybdenum disulphide slabs. Active sites are coordinatively unsaturated molybdenum atoms exposed at the slab edges.

The compounds which are most difficult to remove from petroleum crudes are heterocyclics among which thiophene is typical of S-heterocyclics. How thiophene binds at the active site of a molybdenum disulphide catalyst has been the subject of much research and speculation. There are two extreme possibilities: through the ring, which is then parallel with the catalyst, or through the sulphur atom, when the ring is perpendicular to the surface. It is also conceivable that thiophene molecules may stack on the basal plane of molybdenum disulphide (like benzene on carbon) and on exposed alumina, presumably with the ring parallel with the surface.

In this paper we describe our use of inelastic neutron scattering (INS) to investigate the interaction of thiophene with a γ-alumina supported molybdenum disulphide catalyst. Our aim in this preliminary study was to find out what information the INS technique could yield about the interaction of thiophene with a molybdenum disulphide catalyst. Molybdenum in the catalyst was sulphided but the catalysts were not prereduced in hydrogen. The number of exposed molybdenum atoms is, therefore, likely to be small and we would expect thiophene to interact, if at all, with the basal plane.

Inelastic Neutron Scattering

Inelastic neutron scattering (INS) is a spectroscopic technique which reveals the vibrational, rotational, torsional and translational motions of atoms in catalysts and sorbed species.[2] The spectrum, which looks like an infrared spectrum, is an energy-loss spectrum due to energy transfer from the incident neutrons to the scatterer. Since hydrogen is the strongest scatterer the technique is especially useful for investigating motions associated with hydrogen atoms. Neutron scattering intensities are straightforwardly related to atomic displacements (eigenvectors) and to the concentration of the scattter.

Neutron scattering is described by a scattering law, $S(Q,\omega)$, which connects the momentum transfer, Q, and the energy transfer, ω, with the mean square displacement of the oscillator, U^2:

$$S_n(Q,\omega) = (1/n!)(Q^2U^2)^n \exp(-Q^2U^2)$$

where n is an integer (1 for the fundamental, 2 for the first overtone, etc); ω is the oscillator frequency; U^2 equals $h/2\mu\omega_0$ where μ is the oscillator reduced mass. The spectrum is displayed as the scattering law vs the energy transfer.

2 EXPERIMENTAL

Materials

Molybdenum/alumina catalysts in the oxide form were prepared by the pore-filling method using ammonium heptamolybdate and calcined at 770 K for 2 h. Molybdenum loadings were 8% and 14% MoO_3 and surface areas 240 m^2g^{-1}. The catalysts were sulphided in a H_2/H_2S mixture (10:1) at 625 K for 2 h. The sulphided catalysts were sealed in glass containers under nitrogen.

The containers were opened, and the catalysts handled, in a glove bag filled with argon. The catalysts were dosed with thiophene, liquid or vapour, at room temperature and pumped under vacuum to constant weight. The dosed materials, ca 5 g wrapped in aluminium foil, were attached to the instrument probe in the glove bag and plunged into liquid nitrogen for storage prior to transfer to the spectrometer. Materials are designated TX/MoYS/Al where X is the mass of thiophene and Y the mass of molybdenum per gram of catalyst and S means fully sulphided.

Neutron Scattering

Inelastic neutron scattering spectra were obtained on the time-focused crystal analyser (TFXA) spectrometer at the ISIS spallation source at the Rutherford Appleton Laboratory.[3] Spectra were recorded at 20 K during 12 h for energy transfers in the range 0 - 500 meV. Spectra were analysed with the GENIE program at the Rutherford Appleton Laboratory.

3 RESULTS AND DISCUSSION

In this paper we report our results with the Mo14S/Al catalyst. Materials are listed in Table 1. The INS

Table 1 Materials[a]

Sample	Mo14S/Al	T/Mo14S/Al	
Thiophene /g/g catalyst	0	0.0952	0.418
Thiophene adsorbed /g/(g catalyst)[b]	0	0.0737	0.0737
Thiophene in pores /g/(g catalyst)[c]	0	0.0215	0.344
In beam catalyst/g	5.27	4.04	3.91
thiophene/g	0	0.384	1.63

[a] Catalyst 14 wt-% Mo on alumina, fully sulphided; surface area 240 m^2g^{-1}; pore volume 0.5 cm^3 (g catalyst)$^{-1}$. [b] Area of thiophene molecule 45.5×10^{-20} m^2. [c] By difference.

spectra, 0-4000 cm^{-1}, of thiophene, the sulphided catalyst and the dosed catalysts were measured. The region of molecular vibrations, 400-1200 cm^{-1}, is expanded in Figure 1. Assignments of the fundamental vibrations of thiophene, derived from the infrared spectrum[4] are given in Table 2. (In addition to molecular vibrations, there are lattice vibrations below 200 cm^{-1} which will be discussed in a later paper.) Note that at the temperature of measurement (20 K) the

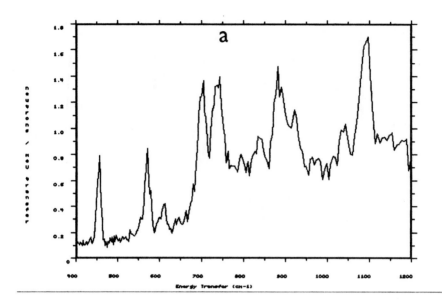

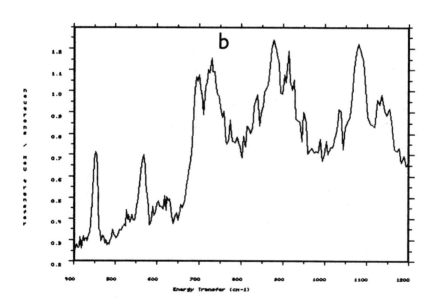

Figure 1   INS spectra in the region of molecular vibrations, 400-1200 cm^{-1}, at 20 K: scattering vs energy transfer (cm^{-1}). (a) pure (solid) thiophene, (b) thiophene (0.418 g/g catalyst) on sulphided Mo14/alumina.

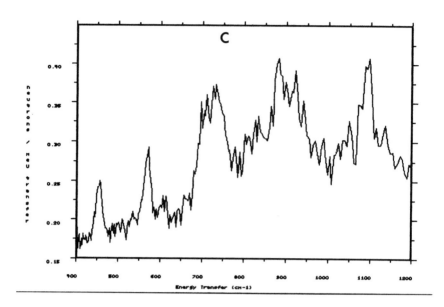

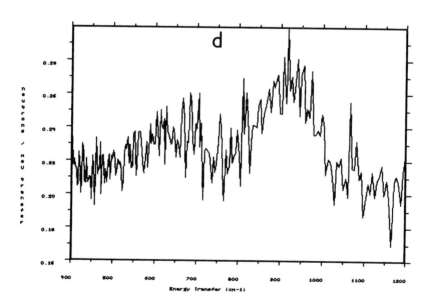

Figure 1 continued (c) thiophene (0.0952 g/g catalyst) on sulphided Mo14/alumina, (d) sulphided Mo14/alumina.

spectrum of thiophene is that of the *solid* (melting temperature 235 K).

Analysis of INS spectra emphasises scattering intensities. These are proportional to the mean square displacement of the oscillator. Intensity changes are due to interaction of the oscillator with another species, in our work, the catalyst. We aim to identify those motions, if any, of the thiophene molecule which are modified in the adsorbed molecule compared with the free molecule. If, for example, the thiophene molecules are lying flat on the catalyst, i.e. having the ring parallel with the surface, then we would expect the out-of-plane oscillations to be most affected by the surface interaction.

Differences between the spectra of pure solid thiophene and adsorbed thiophene can be seen when the spectra of Figures 1b and 1c are compared with 1a. The spectrum of the adsorbed thiophene is broadened and enhanced near 700 and 900 cm^{-1} and in the 450-600 cm^{-1} region. These are the regions of out-of-plane C-H bending (γ-CH) and out-of-plane ring bending (γ-R) modes (see Table 2), precisely the modes likely to be modified when thiophene is lying flat on the catalyst surface. In addition the low wavenumber lattice modes of solid thiophene are suppressed in the adsorbed thiophene. Therefore the spectra of the adsorbed thiophene reveal changes compared with free thiophene consistent with the thiophene molecules lying flat on the catalyst surface.

We proceed with our analysis of the spectra by making use of the integrated intensities in characteristic regions; these are listed in Table 2.

Table 2   Scattering intensities[a]

Region/ cm^{-1}	Assign	Sample and intensities/ neutron counts			
		Thiophene	T.0952/ Mo14S/Al	T.418/ Mo14S/Al	Mo14S/ Al
425-475	γ-R	13.7	10.2	21.1	10.7
650-715	γ-CH	44.7	17.5	48.1	15.4
715-900	δ-R, ν-R	173	59.7	174	43.6
900-950	γ-CH	48.7	17.6	48.7	13.3
1050-1150	δ-CH	112	32.8	97.5	21.0

[a] Samples and sample masses as in Table 1. Assignments: γ, out-of-plane vibration, δ, in-plane bend, ν, stretch; R, ring.

The scattering of the sulphided catalyst is included since it contributes to the overall intensity of the spectra.

The spectra were first normalised to unit mass of *catalyst* in the neutron beam: each sample was weighed before being placed in the neutron beam; the mass of thiophene (known from the mass adsorbed) was subtracted from the sample mass to give the mass of catalyst; the intensities were then divided by the mass of catalyst to give the intensities per unit mass of catalyst in the beam, I_m. Then intensity changes in particular regions of the spectrum. relative to thiophene, ΔI, were calculated from the following expression:

$$\Delta I = \{[I_m - I_m(cat)]/[I_{m,v} - I_{m,v}(cat)] - I(T)/I_v(T)\} / [I(T)/I_v(T)]$$

where $I_m(cat)$ is the intensity from the catalyst, $I_{m,v}(cat)$ is the intensity from the catalyst at the reference wavenumber v, and T refers to pure thiophene. Here v is the region 1050-1150 cm^{-1} (δ-CH). Values of ΔI are given in Table 3.

Table 3 Changes in the thiophene spectra (Δ/%) due to interaction of thiophene with the catalyst

Spectral region /cm-1	Assignment	Intensity changes, Δ/%	
		T.0952/ Mo14S/Al	T.418/ Mo14S/Al
425-475	γ-R	-1.39	31.6
650-715	γ-CH	-14.9	12.0
715-900	δ-R, v-R	1.89	11.8
900-950	γ-CH	2.03	9.57

a Samples as in Table 1. Assignments: γ, out-of-plane vibration, δ, in-plane bend, v, stretch; R, ring.

The results in Table 3 are most striking. In the low thiophene material, T.0952/Mo14S/Al, the only significant change in the spectrum is the intensity decrease at 650-715 cm^{-1} (out-of-plane CH bend), a consequence, presumably, of the thiophene ring interacting with the catalyst surface. In the higher thiophene material the intensity is enhanced especially for the out-of-plane ring bending vibration. For this material only about 20% of the thiophene can be accommodated in a surface monolayer (cf. Table 1). Therefore the spectrum is due to multilayer or pore-entrained thiophene. That the intensities are enhanced compared with free (solid) thiophene means that the effect of the interaction of thiophene with the catalyst (whether in a multilayer or in the pores) is to release the thiophene molecules from the constraints imposed by the lattice in the bulk solid. The amplitudes of the vibrations increase and consequently so do the intensities.

4 CONCLUSIONS

With inelastic neutron scattering we are able to observe the spectrum of thiophene adsorbed at different concentrations on a sulphided Mo/alumina catalyst. Compared with free thiophene the spectrum of adsorbed thiophene shows an intensity loss in the spectral region associated with a CH out-of-plane vibration. Evidently, the adsorbed thiophene molecules are lying flat on the catalyst surface. Entrained thiophene shows intensity enhancements compared with free (solid) thiophene an effect which we attribute to breakdown of the constraints imposed by the lattice in the bulk solid.

REFERENCES

1. P.C.H. Mitchell, in *Catalysis* (Specialist Periodical Report), ed. C. Kemball (The Chemical Society, London 1977), vol. 1, p. 204; P.C.H. Mitchell, in *Catalysis* (Specialist Periodical Report), ed. C. Kemball and D.A. Dowden (The Chemical Society, London 1982), vol. 4, p. 175; R. Prins, V.H.J. de Beer, and G.A. Somorjai, Catal. Rev.-Sci.Eng., 1989, 31, 1.
2. P.C.H. Mitchell and J. Tomkinson, Catalysis Today, 1991, 9, 227; P.N. Jones, E. Knozinger, W. Langel, R.B. Moyes, and J. Tomkinson, Surface Science, 1988, 207, 159; H. Jobic, J. Catal., 1991, 131, 289.
3. J. Penfold and J. Tomkinson, Rutherford Appleton Laboratory Report, RAL-85-050, Chilton, 1985.
4. M. Rico, J.M. Orza, and J. Morcillo, Spectrochim. Acta, 1965, 21, 689; D.W. Scott, J. Mol. Spec., 1969, 31, 451.

The Surface Chemistry of Ceria Doped with Lanthanide Oxides

P. G. Harrison[1], D. A. Creaser[1], B. A. Wolfindale[2], K. C. Waugh[2], M. A. Morris[2], and W. C. Mackrodt[3]

[1] DEPARTMENT OF CHEMISTRY, UNIVERSITY OF NOTTINGHAM, UNIVERSITY PARK, NOTTINGHAM NG7 2RD, UK
[2] ICI CHEMICALS & POLYMERS LTD, KATALCO R & T GROUP, PO BOX I, BILLINGHAM, CLEVELAND TS23 ILB, UK
[3] ICI CHEMICALS & POLYMERS LTD, RESEARCH & TECHNOLOGY DEPARTMENT, PO BOX 13, THE HEATH, RUNCORN WA7 4QF, UK

ABSTRACT

Trivalent cation doping of a host CeO_2 lattice has been used to increase electrical conduction in this type of material. Such doping creates defects within the lattice, the accommodation of the lower valent cation resulting in the creation of anion vacancies and trivalent cerium ions. The defective surface of these materials may enhance the catalytic properties of these solids relative to pure ceria.

Using X-ray photoelectron spectroscopy (XPS), we have studied the surface of ceria doped with various other rare earth oxides. These have been prepared by coprecipitation, calcined at 450°C and then aged at various temperatures in air. It was found that all of the trivalent cations displayed a tendency to segregate to the surface such that a measured surface coverage of up to two monolayers was observed after aging at 1450°C for 24 hours. Finally, we discuss the Ce 3d photoelectron spectra for these materials. It was observed that, during the increase of the lower valent dopant at the surface with temperature, the fraction of three valent cerium cations increased in the surface analyte. This provides strong evidence for the defect mechanism proposed for this type of solid.

INTRODUCTION

Lanthanide oxides have been shown to be active catalysts for a number of commercially important processes including methane coupling[1], butane oxidation[2] and cracking[3]. The activity of the pure oxide materials can be improved by mixing with alternative cations. For this reason, perovskite[4], transition metal lanthanide oxide systems[5], and alkaline earth doped lanthanides[6] have all been investigated as active catalytic materials. The reason for the inherent activity of lanthanide and doped lanthanide materials has been related to the presence of interstitial oxide[6], the participation of lattice oxygen atoms[7], and the basicity of the surface[8]. However different these mechanisms appear, it is clear that the reactivity of these materials is related to the defect chemistry at their surface. Many of the lanthanides can exist in multiple oxidation states, (eg CeO_2, Pr_6O_{11}) and it is the redox cycle possible in these materials (eg $e^- + Ce^{4+} \rightleftharpoons Ce^{3+}$) is involved directly in the reactivity, as has been pointed out by various authors[2,7]. The acid-base properties[8] and the availability of various surface oxygen species[1] are, of course, related to the surface redox reaction.

In order to probe the surface chemistry which may occur in rare earth redox systems, we have used X-ray photoelectron spectroscopy to study ceria doped with various alternative lanthanide oxides. This technique allows surface composition and oxidation tates to be readily determined. The choice of system has been made because the conductivity of ceria can be maximised by doping with trivalent lanthanides[9], and these materials have practical applications in solid oxide fuel cells[10]. Since they have varying electrical[9] and hydrogen reduction properties (related to their efficiency in fuel cells[10]), it is probable that their surfaces exhibit complex redox properties.

EXPERIMENTAL

Samples were prepared of ceria with a 15% dopant level of La^{3+}, Pr^{3+}, Nd^{3+}, Sm^{3+} and Gd^{3+}. The mixed ceria-lanthanide oxide materials were prepared by precipitation from nitrate solutions and were dried at 110°C for 24 hours. About 0.5g of dried sample was calcined in air at 450, 700, 950, 1200, and 1450°C for 24 hours. The resulting oxide was sieved and those particles with a size ca. 1-2mm in diameter were presented for XPS analysis. No further grinding took place prior to analysis. As expected from previous work the XRD patterns from these samples at elevated temperatures showed lines essentially unchanged from those of pure CeO_2. There were small shifts in diffraction angles which result from solid solution of the dopants in the fluorite lattice. Typical data are represented in Figure 1 where 15% La doped ceria is compared with pure ceria.

Experimental data were collected on a VSW multi-technique X-ray photoelectron spectrometer. The background vacuum was better than 5x 10^{-8}mbar during analysis. Samples were mounted on to double-sided tape and evacuated overnight. A high transparency earthed stainless steel gauze was spot-welded over the samples to prevent spillage and to provide uniform sample changing. The samples were evacuated overnight in a reparation chamber prior to entry into the analysis section. For analysis, the samples were irradiated with Al Kα X-rays and photoejected electrons were energy analysed with a hemispherical analyser operating with a pass energy of 50 eV. When charge correction was applied, all quoted binding energies are referenced to an adventitious C 1s signal at 285.0 eV. Data from lanthanide 3d, C 1s and O 1s spectral regions were collected.

RESULTS
1. <u>Surface segregation of trivalent lanthanide ions in ceria</u>.

Typical data sets of cerium 3d and lanthanide 3d core spectra as a function of calcination temperature are shown in Figure 2. The data are illustrated by using spectra collected from a 15% Pr-CeO_2 sample. Both the Ce (Figure 2(a)) and Pr (Figure 2(b)) spectra are more complex than the simple spin-orbit doublet expected due to final state effects where O2p $\longrightarrow$ Ce 4f electron exchange can take place. These are discussed in detail elsewhere[11]. Changes in the form of the spectr are discussed below.

If Figure 2(a) is examined, the region close to the highest binding energy of 938.5eV is the Pr $3d_{5/2}$ feature, shown separately in Figure 2(b). It is striking that, as the calcination temperature increases, this feature is seen to increase relative to the Ce 3d core levels at lower binding energies. At 1450°C this feature now provides the dominant contribution to the spectrum.

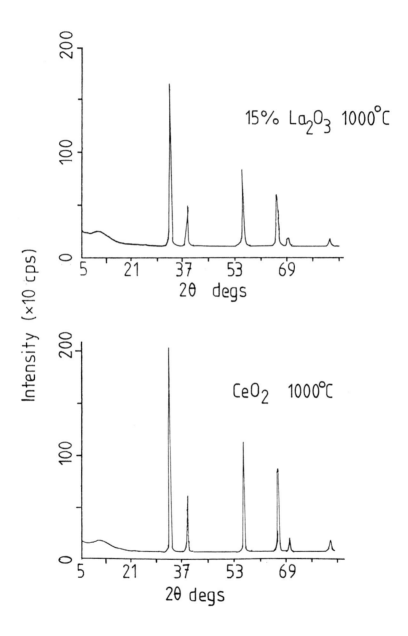

Figure 1. Typical XRD diffractograms from (lower) a ceria powder after 100°C calcination in air and (upper) a 15% La2O3 doped ceria powder after a similar heat treatment.

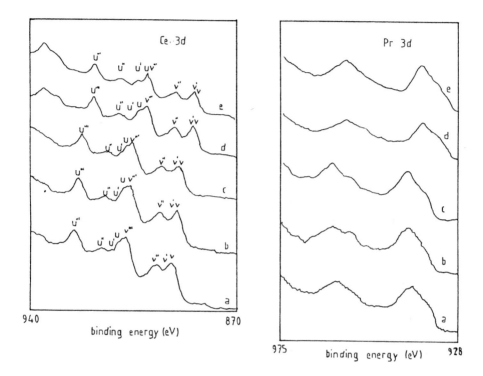

Figure 2. Ce and Pr 3d core level XPS spectra through a cycle of air calcinations: 24 hours at (a) 450, (b) 700, (c) 950, (d) 1200, and (e) 1450°C. Explanation of the features is given in the text.

Similar effects were noted for all the dopant ions investigated. Plots of the lanthanide/cerium total peak area ratio, measured following a Shirley type background subtraction of each photoelectron feature[12], for each lanthanide dopant against temperature are shown in Figure 3. Apart from the Nd and Gd doped systems, there is a gradual increase in the ratio as a function of temperature. The changes which occur at 700°C for Nd and Gd doping may arise from complex solid state reactions. It appears from these data that the lanthanides have a tendency to surface segregate as a function of temperature. Using calculated photoionization cross-sections and escape depths[14] and correct angular terms for the analyser[15], surface coverages of the lanthanide at the ceria surface can be calculated. These are quoted as monolayers relative to the ceria atomic density. Surface coverage dependence with annealing temperature is plotted in Figure 4 for La, Pr and Nd samples showing that up to two monolayers of the lanthanide ion segregate at 1450°C. These data can be converted into Arrhenius type plots by plotting ln(coverage) versus reciprocal temperature, as shown in Figure 5, which yield activation energies (E_{seg}) for segregation in the range 25-45 kJ mol^{-1} from the slopes of these plots.

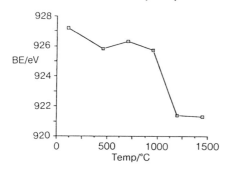
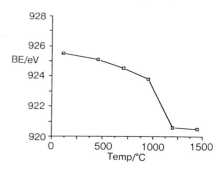
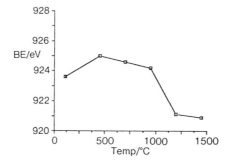
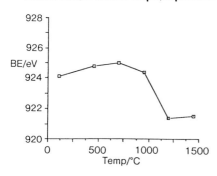
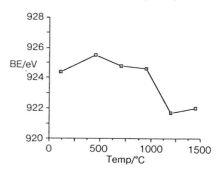

Figure 3. Ln/Ce 3d core level peak area ratio for the doped ceria powders against temperature of calcination. Peak areas were estimated following a Shirley type background subtraction.

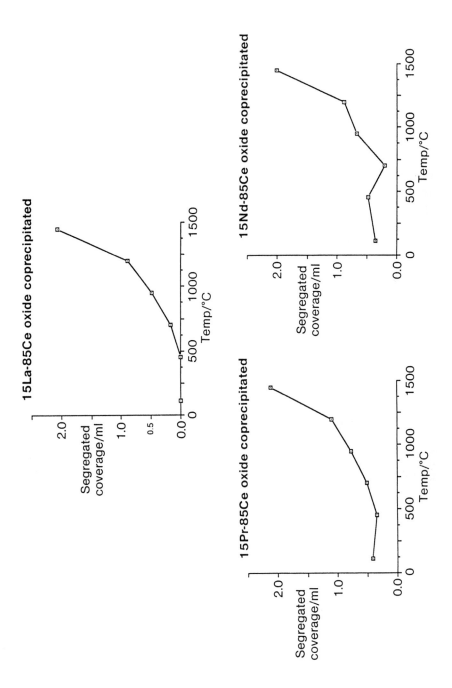

Figure 4. La, Pr and Nd coverages for these doped cerias against calcination temperature. Coverages are quoted as monolayers relative to the number of cations in the ceria surface.

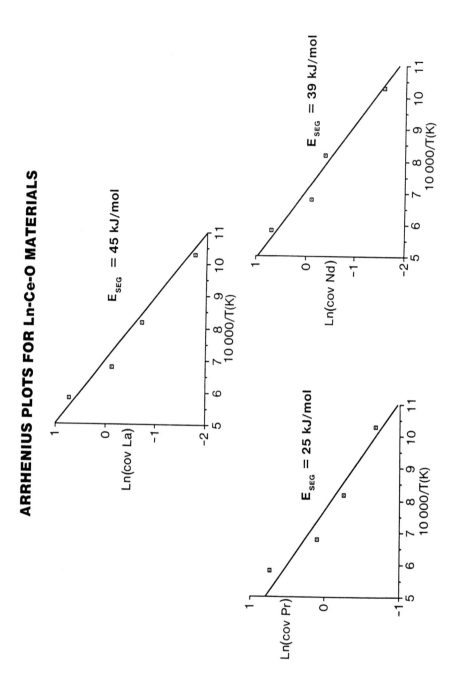

Figure 5. Data given in Figure 4 plotted as an Arrhenius type plot: ln(surface coverage) *versus* reciprocal temperature. Derived activation energies (E_{seg}) are shown and a discussion of these is given in the text.

2. Spectral Changes on Segregation of the Dopant.

Figure 2(a) displays the Ce spectrum through the calcination treatments. As the calcination temperature increases there are distinct changes in the Ce 3d spectra. The main spectral features have been assigned according to convention[16]: u and v represent the two sets of spin orbit multiplets ($3d_{3/2}$ and $3d_{5/2}$, respectively); u^{III} and v^{III} are trhe $3d^9\ 4f^0$ photoemission final state whilst v and v^{II} (also u and u^{II}) come from mixing of the $3d^9\ 4f^2$ and $3d^9\ 4f^1$ final states. States v^I and u^I are unique photoelectron features from the Ce^{3+} state which also exhibits features similar to the u and v features indicated ^{II}.

The photoelectron peak changes are difficult to follow because the increasing lanthanide coverage at the surface will affect the binding energy of the cerium core levels. Nevertheless, several changes can be observed. As the calcination temperature increases, the u peak increases as v^{III} decreases. Also, the u^I feature becomes more distinct through the heating cycles. v^I and v features also change and, as the temperature increases, v^I becomes the most dominant of the two features. These changes have been associated with an increase in the amount of Ce^{3+} at the surface as ceria is reduced using *in vacuo* heating or chemical reduction[11].

More obvious than the subtle changes in the complex photoelectron spectrum are changes in the peak shapes and positions. At lower calcination temperatures the photoelectron peaks are broad and only narrow into sharp features. The broadening of the photoelectron peaks arrises from differential charging under X-ray irradiation since the sample is an electrical insulator. The narrowing of the features indicates increased electrical conductivity within the particles. This has been observed in pure ceria samples[11].

The charging of the sample can also be seen in the position of the photoelectron features. As can be readily seen, there is a movement to lower binding energy through the heating cycle (the spectra are not charge corrected). It has been suggested that the measurement of the apparent binding energy of the Ce u^{III} feature can be used to indicate the amount of Ce^{3+} at the surface.

In Figure 6 the binding energy of the Ce u^{III} feature is plotted against calcination temperature. As can be seen for all the systems studied, there is a decrease in binding energy as the temperature increases.

DISCUSSION

Segregation in oxide systems is not unknown[17]. Segregation in the present study arises from the increased size of the lanthanide 3+ ion relative to the host Ce^{4+} ion. Note that prseseodymium oxide is defective and is a mixture of 3+ and 4+ ions. The values measured from the Arrhenius type plots are somewhat difficult to interpret since the system is probably not at equilibrium and the energies determined will not be segregation energies. Rather, the values measured are probably an indication of the activation energy involved in the transport mechanism. The low value of these parameters in the systems measured, 25-45 kJ mol^{-1}, suggest that the transport mechanism is via grain boundary diffusion rather than a bulk diffusion process via ion exchange which would be expected to occur at much higher energies[18].

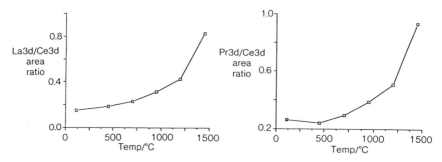

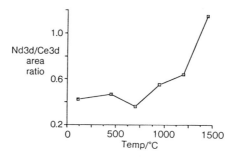

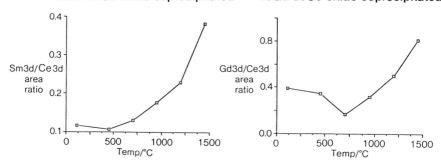

Figure 6. Binding energies of the Ce 3d u^{III} feature against calcination temperature for each of the dopant materials.

We have not considered any complex geometrical factors which could account for the increase in lanthanide ion 3d intensity with increasing temperature (eg growth of the particles decreasing interior/exterior surface area ratio presented to the X-ray beam). Whilst these would affect the signal intensities, they would not be expected to bring about the complex photoelectron peak shape charges seen during calcination to higher temperatures.

The spectral changes are associated with the segregation which is accompanied by reduction of some Ce^{4+} ions in the surface region to Ce^{3+}. This process (ie the doping of lanthanides into ceria) is known to increase the electrical conductivity of ceria[9]. It has been suggested on theoretical grounds[19] that the most likely defect mechanism for ceria is through the production of trivalent cerium ions and anion vacancies. This can be written:

$$O_2 + 4Ce_{Ce}^{4+} + 2V_o^{**} \longrightarrow 4Ce_{Ce}^{4+} + 2O^{2-} \qquad (1)$$

where V_o^{**} is an anion vacancy. This mechanism is consistent with the increase in Ce^{3+} seen here since, as the 3+ lanthanide ions are incorporated into the lattice, charge neutrality must be maintained by loss of oxygen anions (as oxygen) and the equilibrium in equation (1) will move to the left. This is more exactly written:

$$La_2O_3 + 2Ce_{Ce}^* + O_o^* \longrightarrow 2La_{Ce'} + V_o^{**} + 2Ce^{4+} \qquad (2)$$

in standard Kroger-Wink nomenclature[20]. In this way, the results presented herein support the hypothesis that ceria is defective through the production of trivalent cations and anion vacancies. It should be emphasised that the defect nature of a surface will control its chemical reactivity, and that these materials should have markedly different catalytic properties to pure CeO_2.

CONCLUSIONS

XPS spectra of CeO_2 have been collected as the sample has been reduced and oxidised "*in situ*". Detailed analysis and comparison with reference spectra from various compounds show that features in the CeO_2 spectra at lower binding energies derive from "shake-off" from the $3d^9\ 4f^0$ core hole state. These correspond approximately to $3d^9\ 4f^2$ (the extra f electrons arising from O2p levels) states. The observed $3d^9\ 4f^2$ feature is a mixture of $3d^9\ 4f^2$ and $3d^9\ 4f^1$ final states. The observed other 4f final state ($3d^9 4f^1$) is at an apparent binding energy different to that observed in Ce_2O_3 spectra. The Ce_2O_3/CeO_2 ratio observed at the surface of the sample strongly affects the absolute binding energy positions of the Ce 3d features. The increase in conductivity as the sample is reduced (Ce_2O_3 enhanced) is reflected in lower charging of the sample and a concomitant reduction in binding positions and sharpening of the peaks.

ACKNOWLEDGEMENTS

One of us (D.A.C.) would like to thank the SERC and ICI plc for a CASE studentship. The authors would like to thank M. Fowles and M. Gray for all their practical help and invaluable advice. The ICI Strategic Research Funding Scheme is acknowledged for project support.

REFERENCES

1. K.D. Campbell, M. Zhang and J.H. Lunsford, J. Phys. Chem., 1988, **92**, 750.
2. T. Hattori, J.-I. Inoko and Y. Murakami, J. Cat., 1976, **42**, 60.
3. T.M. Tri, J. Massardier, P. Gallezot and B. Inelik, Proceedings of the 7th International Congress on Catalysis. Tokyo, T. Seiyama and K. Tanabe (eds), Elsevier, Amsterdam, 1980.
4. R.J.H. Voorhover, "Advanced Materials in Catalysis", Academic Press New York, 1973, p 129.
5. Y. Okomoto, K. Adachi, T. Imanaka and S. Teranishi, Chem. Letters, 1974, 241.
6. Y. Osada, S. Koike, T. Fukushima, S. Ogasawara, T. Shikada and T. Ikariya, Appl. Cat., 1990, **59**, 59.
7. K. Otsuka and M. Kunitomi, J. Cat., 1987, **105**, 525.
8. V.R. Choudhary and V.H. Rane, J. Cat., 1991, **130**, 411.
9. R. Gerhardt and A.S. Nowick, J. Am. Ceram. Soc., 1988, **69**, 641.
10. H. Yahiro, K. Eguchi and H. Arai, Solid State Ionics, 1989, **36**, 71.
11. P.G. Harrison, D.A. Creaser, B.A. Wolfindale, K.C. Waugh, M.A. Morris and W.C. Mackrodt, proceeding paper.
12. D.A. Shirley, Phys. Rev. B, 1972, **5**, 4709.
13. J.H. Schofield, J. Electron Spec., 1976, **8**, 129.
14. M.P. Seah and W.A. Dench, Surf. Interface Anal., 1979, **1**, 2.
15. R.F. Reilman, A. Msezane and T. Manson, J. Elect. Spect., 1976, **8**, 389.
16. P. Burroughs, A. Hamnett, A.F. Orchard and G. Thornton, J. Chem., Soc., Dalton Trans., 1976, 1686.
17. See for example: M. Cotter, S. Campbell, C.C. Cao, R.G. Edgell and W.C. Mackrodt, Surf. Sci., 1989, **208**, 267.
18. R.M. Barrer, "Diffusion In and Through Solids", Cambridge, 1941.
19. M. Hillert and B. Jansson, J. Am. Ceram. Soc., 1986, **69**, 732.
20. See for exmple: O.T. Sorensen, "Non-stoichiometric Oxides", Academic Press, New York, 1981.

FTIR Identification and Microcalorimetric Characterization of Mo=CH$_2$ and (π-C$_2$H$_4$)Mo^{4+} Complexes after Cyclopropane Adsorption on Photoreduced MoO$_3$/SiO$_2$ Olefin Metathesis Catalysts

K. A. Vikulov[1], B. N. Shelimov[1], V. B. Kazansky[1],
G. Martra[2], L. Marchese[2], and S. Coluccia[2]
[1] N. D. ZELINSKY INSTITUTE OF ORGANIC CHEMISTRY,
LENINSKY PROSPECT 47, 117913 MOSCOW, RUSSIA
[2] UNIVERSITÀ DI TORINO, DIPARTIMENTO DI CHIMICA INORGANICA,
CHIMICA FISICA E CHIMICA DEI MATERIALI, VIA P. GIURIA 7,
10125 TORINO, ITALY

SUMMARY

Room temperature photoreduction in CO was used to produce Mo^{4+} centers on the surface of MoO$_3$/SiO$_2$ catalyst after high temperature dehydroxylation and oxidation. It was found previously that cyclopropane adsorption at room temperature increased sharply activity of the photoreduced catalyst in olefin metathesis reaction. FTIR spectroscopy was applied to identify surface species formed after cyclopropane interaction with Mo^{4+} ions. It was found that cyclopropane formed mainly two types of species: Mo=CH$_2$ carbene complexes (absorption bands at 3078±2 and 2945±2 cm^{-1}) and (π-C$_2$H$_4$)Mo^{4+} complexes (3078±2, 3060±2, 2998±2 and 1404±2). C$_2$D$_4$ and ^{13}C$_2$H$_4$ were used to confirm band assignment. Mo=CH$_2$ were found to be thermally stable up to 350°C in vacuum while ethene π-complexes decomposed at this temperature. Heats of chemisorption of cyclopropane (-207±13 kJ/mole) and ethene (-158±8 kJ/mole) on photoreduced catalyst at 40°C were measured using a conventional Calvet-type microcalorimeter. Stoichiometry of chemisorption evidenced that each cyclopropane molecule interacted with two Mo^{4+} centers. Following FTIR results on simultaneous Mo=CH$_2$ and (π-C$_2$H$_4$)Mo^{4+} formation during cyclopropane chemisorption the bond energy of metal-carbene was calculated to be 435±25 kJ/mole.

1 INTRODUCTION

Low temperature photoreduction was recognised as an interesting technique for activation of supported metal oxides[1]. In case of supported MoO$_3$/SiO$_2$ it was shown that after photoreduction in CO at room temperature Mo^{6+} ions were reduced to Mo^{4+} mainly[2]. It was found later that photoreduced sample became a very active catalyst for olefin metathesis reaction$_3$ after cyclopropane chemisorption at room temperature. The effect was considerd to be due to the Mo=CH$_2$ carbene complexes formation (assumed active centers of the reaction):

$$Mo^{4+} + cyclo\text{-}C_3H_6 \rightarrow Mo=CH_2 + C_2H_4 \qquad (1)$$

Mo=CH$_2$ and Mo=CD$_2$ carbene complexes were characterised by 3080(2245) and 2945(2160) cm^{-1} absorption bands in diffuse reflectance IR spectra of chemisorbed cyclopropane$_4$. It was supposed that ethene molecule formed from cyclopropane could react with an uncoordinated Mo^{4+} center giving a π-like complex.

To justify these proposals additional FTIR spectroscopic studies of cyclopropane chemisorption on photoreduced MoO$_3$/SiO$_2$ catalyst were performed in this work. Special attention was paid to search for absorption bands of chemisorbed cyclopropane below 2000 cm^{-1} as SiO$_2$ was completely "dark" in this region with diffuse reflectance technique$_4$.

Microcalorimetric mesurements were applied to obtain heats of cyclopropane and ethene chemisorption on photoreduced catalyst in order to estimate the Mo-C bond energy in Mo=CH$_2$ carbene complexes.

2 EXPERIMENTAL

Industrial KSK-2-5 silica (S$\approx$250-270 m^{2}g^{-1}) was impregnated with an aqueous solution of (NH$_4$)$_6$Mo$_7$O$_{24}$ and dried at room temperature. Concentration of the solution was chosen to give respectively 1 and 2.5 weight % of Mo in the dried sample.

In order to increase the intensity of absorption bands of chemisorbed cyclopropane it was necessary to increase the thickness of a pellet. This caused the decrease of the sample transparency below 2000 cm^{-1}. The compromise was found to work with two pellets of different thickness and accordinly of different transparency in that region. The "thin" pellet was used below 2000 cm^{-1} and the "thick" one - above 2000 cm^{-1}. The "thin" pellet was approximately of 0.02 g cm^{-2} and the "thick" one of 0.1 g cm^{-2}. These pellets were supported in gold frames and put into conventional quarz vacuum cells for IR measurements with KBr windows.

Further pretreatment of the pellets included calcination in oxygen (100 Torr) at 800°C for 1h and also in vacuo at the same temperature for 1h. Then dry O$_2$ was reintroduced into the cell and the system was cooled down to 500°C and evacuated at this temperature for 1h. At this step the pellet was white. After cooling the cell to room temperature dry CO was introduced (about 10 Torr) and photoreduction with unfiltered light of a high pressure mercury lamp (500 Wt) was performed over 2 or 3 hs. A spherical mirror was used to condense the flux on the pellet. CO_2 was permanently trapped with liquid N_2 during photoreduction. CO was outgassed at room temperature and then the sample was heated at 150°C to evacuate chemisorbed CO. After such treatment the pellets were bright green in color, characteristic of uncoordinated Mo^{4+} ions on silica[2,4].

The cell containing the photoreduced sample was transferred to a BRUKER-48 FTIR instrument and connected to a vacuum line allowing all subsequent adsorption-desorption experiments to be carried out *in situ*. The spectra were recorded with 4.0 cm^{-1} resolution and 4000 accumulating scans.

For microcalorimetric measurments the powder of 1% Mo/SiO_2 catalyst ($\approx$2g) was subjected to the same pretreatment and photoreduction procedure in a quartz reactor with circulating flow of gases. The number of Mo^{4+} centers ($N(Mo^{4+})$) formed after photoreduction was calculated after measuring the amount of CO_2 in the trap which was about $3*10^{19}$ g^{-1} corresponding to $\approx$50% of the total number of Mo atoms in the sample. The powder finally was divided under vacuum into small portions of 0.1-0.2 g each and transferred inside breakable glass bulls.

The heats were measured with an automatic Calvet-type microcalorimeter thermostated at 40°C. Small doses of gas $(2-4)*10^{17}$ molecules were sent to the sample in each run. The heat, calculated after complete adsorption of the dose and achievement of equilibrium (after about 1h), was related to the number of adsorbed molecules. The relative uncertainty of the experiments concerning calibration of volumes, microcalorimeter and pirani gauge was about ±5%.

"Matheson" high purity grade O_2, CO, cyclopropane, C_2H_4, C_2D_4 and $^{13}C_2H_4$ were used in this study. Hydrocarbons were purified by repeated freeze-pump-thaw cycles before each run.

3 RESULTS AND DISCUSSION

FTIR Results

Small doses of cyclopropane were adsorbed up to the saturation of all Mo^{4+} centers in the sample to avoid side reactions under the excess of the gas. Three curves in Fig.1 (plots 1, 2 and 3) represent changes in the spectrum of photoreduced catalyst during this treatment

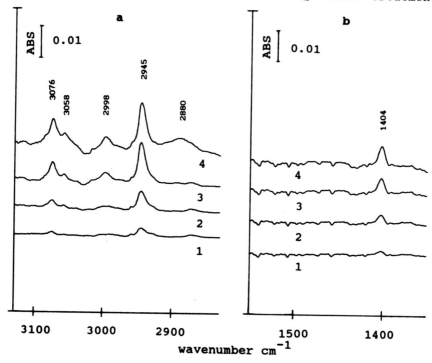

Figure 1 FTIR spectra following cyclopropane adsorption on photoreduced MoO_3/SiO_2 catalyst. Plots 1-3: spectra obtained upon admission of successive doses of cyclopropane up to saturation of Mo^{4+} ions; plot 4: after prolonged contact with excess cyclopropane.

in 3130-2830 (a) and 1550-1350 (b) cm^{-1} regions. Five absorption bands can be identified in the spectrum of chemisorbed cyclopropane at room temperature: 3076±2, 3058±2, 2998±2, 2945±2 and 1404±2 cm^{-1}. Keeping the sample in the excess of cyclopropane for 10 min at room temperature caused some changes in the spectrum i.e. a new absorption band about 2880 cm^{-1} appeared in Fig.1 (Plot 4). The last spectrum containing four intense absorption bands at 3076, 2998, 2945 and 2880 cm^{-1} is quite similar to that measured with diffuse reflectance IR technique[4].

Evacuation at room temperature did not give any change in the FTIR spectrun of chemisorbed cyclopropane. After heating in vacuum at 350°C the spectrum was simplified as the absorption bands at 3058, 2998 and 1404 cm^{-1} disappeared, the intensity of the 3076 cm^{-1}

absorption band was reduced whereas the absorption band

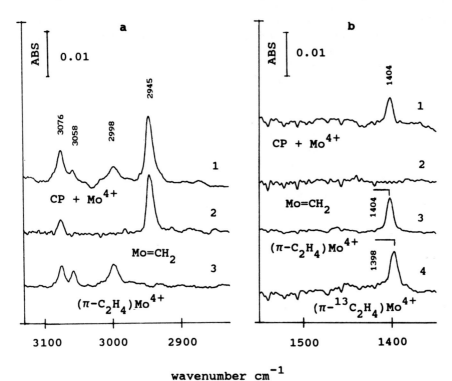

Figure 2 IR evidence for Mo=CH_2 and $(\pi-C_2H_4)Mo^{4+}$ formation when cyclopropane is adsorbed on photoreduced MoO_3/SiO_2 catalyst. Plot 1: cyclopropane + Mo^{4+}; plot 2: Mo=CH_2; plot 3: $(\pi-C_2H_4)Mo^{4+}$; plot 4: $(\pi-^{13}C_2H_4)Mo^{4+}$.

at 2945 cm^{-1} did not change. Apparently two AB overlap at 3076 cm^{-1}; one was removed and the other not removed by evacuation at 350°C. Overall two types of adsorbed species are identified associated with two sets of absorption bands at 3076, 3058, 2998 and 1404 cm^{-1} and at 3076 and 2945 cm^{-1}. The latter couple of absorption bands were previously assigned to the asymmetric and symmetric CH stretching modes in Mo=CH_2 wich was not desorbed by evacuation at 350°C. The former set of four absorption bands might be due to some less stable product of cyclopropane interaction with Mo^{4+} ions than Mo=CH_2 complexes. It was assumed that this product was a π-like ethene complex with Mo^{4+} ion. To confirm this assignment FTIR study of ethene chemisorption on the photoreduced catalyst was performed.

Plot 3 in Fig.2 represents the FTIR spectrum of ethene chemisorbed on Mo^{4+} ions up to their saturation. Comparison of spectra 1, 2 and 3 on Fig.2 evidences that ethene complexes have the same absorption bands as those formed after cyclopropane chemisorption and disappeared after evacuation at 350°C (3076±2, 3058±2, 2998±2 and 1404±2 cm^{-1}) so that spectrum 1 is the sum of spectra 2 and 3. According to the IR bands of some ethene π-complexes described in literature[5-8] it is possible to assume that absorption bands at 3076, 3058 and 2998 cm^{-1} correspond to three of four CH stretching vibrations of π-bonded ethene. Absorption bands at 1404 cm^{-1} may be attributed to both C=C double bond stretching and CH_2 bending modes in these complexes. In order to clarify the assignment, experiments using C_2D_4 and $^{13}C_2H_4$ were performed.

In the case of C_2D_4 no band was found between 1800 and 1350 cm^{-1}. Chemisorbed $^{13}C_2H_4$ showed an intense absorption band about 1398±2 cm^{-1} presented in Fig.2 (plot 4). Assuming that carbon atom labelling does not affect significantly the frequency of a bending mode the observed absorption band of chemisorbed ethene at 1404 cm^{-1} must be assigned to one of the CH_2-group deformations. Due to low transparency of silica below 1350 cm^{-1} it was not possible to observe other absorption bands of these complexes.

Comparing IR spectra of cyclopropane chemisorbed on photoreduced MoO_3/SiO_2 catalyst on Fig.1 and 2 with those published in the first contribution[4] one can see that the results are reproducible. It seems likely that in the first diffuse reflectance studies absorption bands near 3076 and 3058 cm^{-1} were not resolved and appeared as one band near $\approx$ 3080 cm^{-1} (as the resolution was not better than 10 cm^{-1}). An additional absorption band near 2870 cm^{-1}, found with diffuse reflectance, was observed also in this study near 2880 cm^{-1} when excess of cyclopropane was allowed onto the catalyst. It is possible to propose that this band is due to CH_3-group stretching vibration of propene π-complexes formed during secondary reactions of cyclopropane with the surface because traces of propene were found previously in thermodesorption products[4] of chemisorbed cyclopropane.

Microcalorimetric Results

The dependence of the cyclopropane heat of chemisorption on the coverage of Mo^{4+} centers Θ = $N(CP)/N(Mo^{4+})$ is presented in Fig.3 (plot A). The heat was constant (-207±12 kJ/mole) for Θ ≤ 0.5. It decreased sharply at Θ ≈ 0.5 and new doses of cyclopropane were not adsorbed any more. After outgassing cyclopropane at 20°C and then at 350°C the sample became able to adsorb it again with approximately the same heat (plot B). This time the adsorption stopped at Θ about 0.75. The same

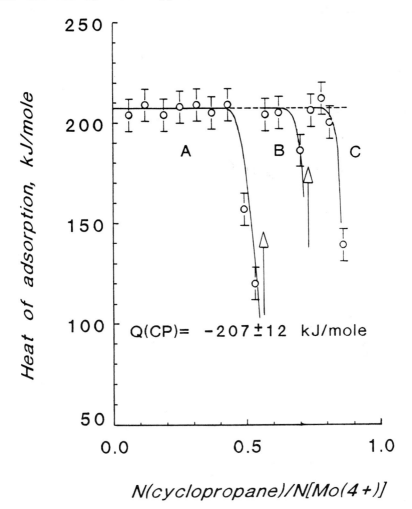

Figure 3 Isosteric heat of CP successive doses adsorption on photoreduced MoO_3/SiO_2 (for plots A, B and C see description in the text).

phenomenon was observed after repeating the cycle (plot C) and finally coverage as high as Θ ≈ 0.9 was obtained.

Following FTIR results these experiments indicate that each cyclopropane molecule reacts with two Mo^{4+} centers one forming a thermally stable $Mo=CH_2$ and the other a less stable ethene complex. That is why adsorption stops when Θ approaches 0.5 (plot A). After

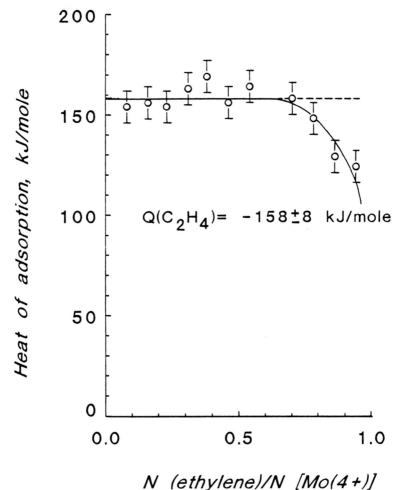

Figure 4 Isosteric heat of ethene successive doses adsorption on photoreduced MoO_3/SiO_2.

heating at 330 °C ethene complexes decompose giving uncoordinated Mo^{4+} centers their amount being about 50% of that in original sample. During the second chemisorption 50% of these restored centers are converted into $Mo=CH_2$ and 50% to ethene complexes (plot B) so that the sample has now 75% of initial number of Mo^{4+} ions in carbene complexes and 25% in π-complexes. Repeating the procedure (plot C) leads to conversion of about 90% of Mo^{4+} ions to $Mo=CH_2$.

The heat of ethene adsorption was constant (-158 ± 8 kJ/mole) up to $\Theta \approx 0.8$ and then started decreasing (Fig.4) indicating that ethene tended to react with Mo^{4+} ions with stoichiometry close to 1:1.

Evaluation of the Mo-C Bonding Energy in $Mo=CH_2$

Estimation of the bond energy metal-carbon in transition metal carbene complexes is of great theoretical and practical importance as these complexes are involved in a number of organic and catalytic reactions[9]. Based on mass-spectroscopic data Beauchamp et al.[10] obtained values of bond strengths in gas phase $M-CH_2^+$ ions for various transition metals but not for Mo. For $W-CH_2^+$ and $Re-CH_2^+$ (supported oxides of these metals together with Mo are considered to be the most active olefin metathesis catalysts) they inferred values lower than 111 kcal/mole (465 kJ/mole) and 102 ± 9 kcal/mole (427 ± 38 kJ/mole) respectively. Using quantum chemical calculations Rappe and Goddard[11] found the bond strength of Mo-C in oxochloride complex $(Cl)_2(O)Mo=CH_2$ to be 297 kJ/mole.

The microcalorimetric results of this study enable us to estimate the bond energy in $Mo=CH_2$ complexes on the surface of a real olefin metathesis catalyst. Using tabulated values of $\Delta H_f^\circ(:CH_2, {}^3B_1) = 390 \pm 4$ kJ/mole, $\Delta H_f^\circ(cyclopropane) = 53.3 \pm 0.6$ kJ/mole and $\Delta H_f^\circ(C_2H_4) = 52.5 \pm 0.4$ kJ/mole[12] and applying the relation:

$$\Delta U_r (const.volume) = \Delta H_r (const.pressure) - \Delta nRT \quad (2)$$

it is possible to give theoretical value of $\Delta U_1 = 386 \pm 5$ kJ/mole for the following reaction in gas phase:

$$cyclo-C_3H_6 \rightarrow :CH_2({}^3B_1) + C_2H_4 \quad (3)$$

Our experimental results give $\Delta U_2 = -207 \pm 12$ and $\Delta U_3 = -158 \pm 8$ kJ/mole for two reactions respectivly:

$$cyclo-C_3H_6 + 2Mo^{4+} \rightarrow Mo=CH_2 + (\pi-C_2H_4)Mo^{4+} \quad (4)$$

$$C_2H_4 + Mo^{4+} \rightarrow (\pi-C_2H_4)Mo^{4+} \quad (5)$$

Using these data it is possible to estimate $\Delta U_4 = \Delta U_2 - \Delta U_3 - \Delta U_1 = -435 \pm 25$ kJ/mole for the reaction:

$$:CH_2({}^3B_1) + Mo^{4+} \rightarrow Mo=CH_2 \quad (6)$$

This value is in agreement with the prediction of Stivens and Beauchamp that for an active olefin metathesis catalyst the metal-carbene bonding energy must be higher than 100 kcal/mole (419 kJ/mole)[13] but is significantly higher than the theoretical result (-297 kJ/mole) calculated by Goddard[11].

In conclusion the results of this study give an additional strong support to the proposed mechanism of cyclopopane interaction with Mo^{4+} ions on silica[4] (reaction 4) and to the value of Mo=C double bond energy (435±25 kJ/mole) in $Mo=CH_2$ carbene complexes calculated under the assumption of this mechanism.

ACKNOWLEDGMENTS

Financial assistance by CNR and ASSTP is gratefully acknowledged.

REFERENCES

1. V.B.Kazansky, Kinetics and Catalysis, 1983,24,1135.
2. B.N.Shelimov, A.N.Pershin and V.B.Kazansky, J.Catal., 1980,64,426; I.V.Elev, B.N.Shelimov and V.B.Kazansky, J.Catal., 1988,113,256.
3. I.V.Elev, B.N.Shelimov and V.B.Kazansky, Kinetics and Catalysis, 1989,30,787.
4. K.A.Vikulov, I.V.Elev, B.N.Shelimov and V.B.Kazansky, J.Mol.Catal., 1989,55,126.
5. J.Hiraishi, Spectrochim.Acta, 1969,25A,749.
6. D.C.Andrews and G.Davidson, J.Organometal.Chem., 1972,36,161.
7. M.Bigorgne, J.Organometal.Chem., 1978,160,345.
8. J.A.Crayston and G.Davidson, Spectrochim.Acta, 1986, 42A,1385; ibid, 1987,43A,559.
9. K.H.Dötz and E.O.Fisher, "Transition Metal Carbene Complexes", Verlag Chemie, 1983.
10. K.K.Irikura and J.L.Beauchamp, J.Chem.Phys., 1991, 95,8344.
11. A.K.Rappe and W.A.Goddard, J.Amer.Chem.Soc., 1980, 102,5114; ibid, 1982,104,448.
12. J.B.Pedley, R.D.Naylor and S.P.Kirby, "Theoretical Data of Organic Compounds", Chapman and Hall, London-New York,1986.
13. A.E.Stivens and J.L.Beauchamp, J.Amer.Chem.Soc., 1979,101,6449.

Infrared Spectroscopic Studies of Chemisorption on Metal Surfaces

M. A. Chesters[1], P. Hollins[2], and D. A. Slater[2]
[1] SCHOOL OF CHEMICAL SCIENCES, UNIVERSITY OF EAST ANGLIA, NORWICH NR4 7TJ, UK
[2] DEPARTMENT OF CHEMISTRY, UNIVERSITY OF READING, WHITEKNIGHTS, PO BOX 224, READING RG6 2AD, UK

1 INTRODUCTION

Vibrational spectroscopy, and particularly infrared spectroscopy, is a powerful and versatile technique for qualitative analysis which finds applications across the complete range of sample phases. Infrared spectroscopy has been successfully applied to adsorbates on high surface area materials, such as catalysts, since the 1950s but is only recently finding wide application in studies of adsorption on well-characterised single-crystal surfaces. The ability to operate in conditions varying from ultra-high vacuum (UHV) to high pressure (10's of bar) and to be applied to both low area single-crystal substrates and high area model catalysts makes infrared spectroscopy an ideal tool to bridge the infamous 'pressure gap' between Surface Science and Catalysis research.

This article will examine aspects of recent research using reflection-absorption infrared spectroscopy (RAIRS)* which are particularly relevant to studies of catalysis. Comparisons will be made with parallel investigations using the electron energy loss spectroscopy (EELS) technique which is continuing to make an important contribution in this area. The material in this article does not represent a review of the RAIRS literature but is intended to provide the non-specialist reader with a picture of the contribution the technique is making in research into the fundamental aspects of catalysis.

2 THE RAIRS TECHNIQUE

The optical considerations which govern the design of a RAIRS system suitable for study of adsorption and reactions on single-crystal metal surfaces have been covered in many previous articles, as have the criteria for choice of spectrometer.[1,2] Most current practitioners employ Fourier transform infrared (FTIR) spectrometers which can give simultaneous access to the 4000-400 cm^{-1} region although this range is generally limited by choice of detector. The problem of dealing with the potentially very useful region between 600-100 cm^{-1} (i.e. bridging the mid-infrared and far-infrared regions) is dealt with in section 4.

The parallel light beam from the interferometer is focused at grazing incidence (typically $85 \pm 5°$ from the surface normal) on the sample surface and then refocused at the detector. Multiple reflection is not used as it provides little
*also known as IRRAS

improvement in signal-to-noise ratio (S/N).[1] Since a molecular monolayer on a single crystal surface represents a small sample in IR analysis, typically 10^{-9} mole, S/N performance is vitally important. In recent years a S/N ratio of 100 for a 1% absorption band has been a typical figure of merit but improvements in spectrometer performance will make significant advances on this figure. If we bear in mind that a 1% reflection-absorption band is to be expected from a full monolayer of a strong infrared absorber it is clear that sensitivity will remain a crucial factor in limiting the applicability of the technique.

The Metal Surface Selection Rule

The reason for the grazing angle of incidence on the metal surface is that light reflecting from a perfect conductor may only have an electric amplitude perpendicular to the surface.[3] This is achieved by using p polarized light at grazing incidence. The alternative polarization at right angles to the plane of incidence (s polarization) produces a node in the standing wave at the surface and hence no absorption by a thin surface layer. This effect of the optical properties of the metal leads to the metal surface selection rule which states that only vibrational modes polarized perpendicular to the metal surface may be excited.

This selection rule produces particularly striking results for larger adsorbed molecules (i.e. containing more than two or three atoms). The effect on the spectrum of a thin film of butane adsorbed on a Pt(111) crystal is illustrated in fig.(1), in which the RAIR spectrum is compared to a transmission infrared spectrum of a thin butane film condensed on a potassium bromide plate[4]. The bands which appear in the RAIR spectrum may be assigned to modes polarized perpendicular to the butane carbon skeleton (which is in the all-trans conformation at 100K) while the missing bands, which feature strongly in the spectrum of the film on the KBr plate, may be assigned to modes polarized parallel to the carbon skeleton. This result shows that butane deposits on Pt(111) at this temperature with the carbon skeleton parallel to the surface.

The strict application of the metal surface selection rule is of considerable help in determining adsorbate structure and orientation.

Complementarity to EELS

The EELS technique has been applied to studies of adsorbates on metal surfaces since the mid-1970's.[5,6] Its ability to produce good quality vibrational spectra of hydrocarbon monolayers on metal single-crystal surfaces led to a renaissance of vibrational spectroscopy as a Surface Science technique. While RAIRS was established earlier it remained a specialized technique applicable only to strong infrared absorbers, notably carbon monoxide, until the mid-1980's when Fourier transform spectrometers were first used.

Resolution. The two techniques, RAIRS and EELS, now have similar sensitivities in the 800 to 4000 cm^{-1} range providing absorption bands are not inherently broad (say <20 cm^{-1}). The relatively poor resolution of EELS (30-50 cm^{-1}) is actually an advantage when observing broad bands. For example, the anomalously broad CH stretching band (approximately 300 cm^{-1}) produced by the interaction of cyclohexane and the Pt(111) surface results in a dominant feature in the EEL spectrum but is very difficult to detect by RAIRS.[7,8] The resolution advantage of the infrared technique is particularly valuable in studies of hydrocarbon adsorption where, for instance, the various contributions from the CH stretching modes of methyl and methylene groups of butane shown in fig.(1) would be unresolved in an EEL spectrum. A recent investigation of carbon monoxide adsorption on the Pd(110) surface by RAIRS has shown that high resolution may be necessary even

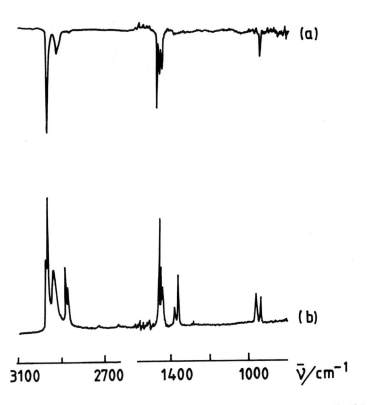

Figure 1 Effect of the metal surface selection rule. (a) RAIR spectrum of a thin film of butane on Pt(111). (b) Transmission infrared spectrum of a thin film of butane on KBr.

for simple adsorbates.[9] A complex series of infrared spectra showing multiple peaks in the intermediate coverage range are believed to arise from reconstruction of the Pd(110) surface under the influence of the adsorbate, fig.(2). At monolayer coverage the surface reverts to its original structure. Earlier spectra recorded using the EELS technique had failed to resolve the multiple peaks and had not led to any suspicion of surface reconstruction.

The above comments apply to virtually all of the EEL spectrometers in current use but there have been recent improvements in resolution. The first 1 meV (8 cm^{-1}) resolution EEL spectrum has been reported by Ibach. If this type of performance becomes widely available the resolution of EEL spectra will become similar to that of RAIRS, which is typically operated at 2-4 cm^{-1}.

Selection Rules. In general, the strong loss features in an EEL spectrum result from excitation by a dipole-scattering mechanism which obeys the same metal surface selection rule as found in RAIRS. However, there are other significant mechanisms resulting from short-range electron-molecule interactions which follow different selection rules and therefore provide valuable additional information on adsorbate structure.[5,6]

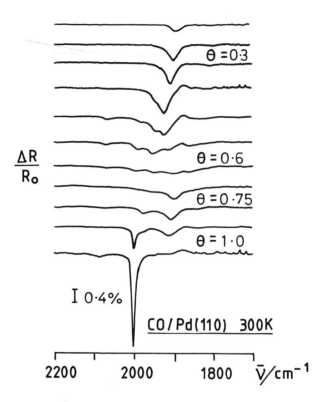

Figure 2 RAIR spectra of carbon monoxide on Pd(110) showing evidence for surface reconstruction. Reprinted from ref (9) with permission.

The low wavenumber region. EELS benefits from a sensitivity advantage in the region below 800 cm^{-1} which increases rapidly at low wavenumbers. This phenomenon results from the greater collection efficiency of low-loss-energy electrons which are 'channeled' close to the specular direction in dipole-scattering because of the small momentum change associated with the energy loss process.[5,6] This advantage has made EELS the technique of choice for study of atomic adsorbates, such as O, S or Cl, and for investigating the low frequency frustrated translations and rotations of adsorbed molecules. The prospects for the application of RAIRS in this important region are covered in section 4.

The Pressure Advantage. Infrared spectroscopy of adsorbed species may be carried out in gas pressures of many bar, providing due care is taken to minimize absorption due to gas phase species. EELS is restricted to high vacuum conditions since operation of any electron source can lead to decomposition of gas molecules and the high resolution of electron spectrometers can be limited by contamination of electrode surfaces. This advantage of the infrared spectroscopic technique is likely to grow in significance as RAIRS is applied more widely to probe reactions at surfaces rather than simple chemisorption. This point will be emphasized in many of the examples which follow.

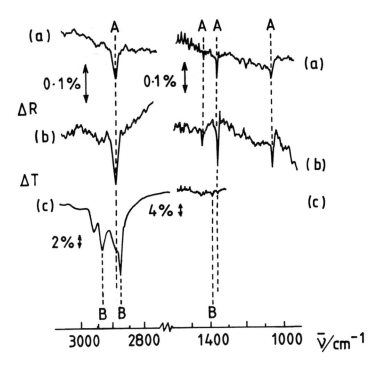

Figure 3 (a) RAIR spectrum of *cis*-but-2-ene adsorbed on Pt(111) at 300K.
(b) RAIRS spectra of but-2-yne adsorbed on Pt(111) at 300K.
(c) Transmission infrared spectrum of *cis*-but-2-ene adsorbed on Pt/SiO$_2$ at 300K.
Bands labelled A are assigned to chemisorbed but-2-yne and bands labelled B are assigned to chemisorbed butylidyne.

3 APPLICATION OF RAIRS TO FUNDAMENTALS OF CATALYSIS

Comparison of IR spectra of Adsorbates on Metal Single Crystal and Supported Metal Catalysts

Supported metal catalysts contain small metal particles (1-100 nm) which expose a variety of adsorption sites on different crystal facets and on steps, edges and corners. It is therefore not surprising that infrared spectra of, for example, adsorbed hydrocarbons tend to be more complex than those of the same adsorbates on single crystal surfaces. Results from the EELS technique provided new insights into the nature of surface species resulting from alkene adsorption through the identification of alkylidyne surface species. Application of RAIRS to this problem has allowed more detailed comparison between supported metal and single crystal results. The contributions from di-σ adsorbed, π-adsorbed and ethylidyne species have each been identified in the transmission infrared spectrum of ethene adsorbed on silica-supported platinum (Pt/SiO$_2$).[11] Correlation between RAIR spectra of propene and but-1-ene adsorbed on Pt(111) at 300K and transmission infrared spectra of these adsorbates on Pt/SiO$_2$ similarly showed propylidyne and butylidyne to be major surface species. It is interesting to note that this correlation broke down in the case of but-2-ene adsorption. The RAIR spectrum of *cis*-but-2-ene adsorbed on Pt(111) at 300K, fig.(3a), shows that the molecule dehydrogenates to but-2-yne (the spectrum of adsorbed but-2-yne is shown for comparison in fig.(3b)). This decomposition route had not been observed for the other alkenes on Pt(111) and presumably arises because butylidyne formation requires the carbon-carbon double bond to be at the end of the carbon chain. In contrast the transmission infrared spectrum of but-2-ene adsorbed on Pt/SiO_2, fig.(3c), shows strong features assigned to butylidyne. It would appear, quite reasonably, that the Pt(111) surface is not a very good model for a metal particle. On the complex catalyst surface the carbon-carbon double bond shift must be able to readily occur, followed by production of the rather stable butylidyne species. This result illustrates the fact that, while RAIRS studies on close-packed single crystal surfaces have provided valuable 'fingerprint' spectra of specific surface species, work on high index and stepped-kinked crystals will be necessary to simulate the range of reactivity of the catalyst. This will need to include work at higher ambient pressure of hydrocarbon and hydrogen.

RAIRS at 'High' Pressure

The 'pressure advantage' of the RAIRS technique is readily appreciated but the technical problems associated with recording very high sensitivity IR spectra in even mbar pressures of ultra-pure gases are considerable. Pritchard and co-workers have described the difficulties arising from low levels of iron carbonyl impurity in 'high purity' carbon monoxide.[12] Once identified this problem was readily overcome and spectra following the reversible, weak chemisorption of carbon monoxide on Cu(100) recorded at ambient pressures up to 30 mbar correlate well with results at low temperature and pressure.[12] The contribution from strong gas-phase absorption was minimized by using s polarized light to record reference spectra in the presence of the adsorbing gas, relying on the fact that the surface layer is 'invisible' to s polarized light.

Reactant pressures in the mbar range are sufficient to initiate catalytic reactions over model catalysts and so RAIRS may be used *in situ* to monitor such processes. Both carbon monoxide oxidation and carbon monoxide hydrogenation have been investigated by F.M. Hoffmann and coworkers.[13,14] In the hydrogenation reaction potassium adsorbed on the Ru(001) substrate was found to promote formation of oxygenated products rather than methane which was the major hydrogenation product on the un-promoted surface. The spectrum of the surface generated under a 2.5 mbar mixture of carbon monoxide and hydrogen, fig.(4a), correlates with the spectrum of a surface potassium formate which is believed to be an important intermediate in the formation of the desired oxygenated products.

Finally we include, using author's licence, an example which qualifies as 'high pressure' only in the sense that spectra were recorded in ambient pressures too high for operation of the EELS technique. Recent work on the very weak chemisorption of ethene on Ag(100) shows the effect of adsorbed chlorine as a promoter in strengthening the silver-ethene bond. The series of spectra in fig.(5) were recorded at increasing equilibrium pressures of ethene over non-promoted Ag(100) at 100K. At pressures up to 1.3×10^{-5} mbar the only band detectable in the spectrum is at 950 cm^{-1} plus a small companion at approximately 960 cm^{-1} associated with surface defects. This band is readily assigned to the strongest i.r. active mode of ethene, v_7, which is an out-of-plane CH deformation, fig.(6). The v_7 mode will generate a strong dynamic dipole moment perpendicular to the silver surface, if the molecule is lying flat, irrespective of any interaction with the surface. The other modes nominally allowed by the metal surface selection rule, v_1 to v_3 in fig.(6), are infrared inactive for the free molecule and generate perpendicular dipoles only through the molecule-metal interaction. The absence of all modes but v_7 in the spectrum is an indication of the very weak nature of the silver-ethene interaction as is the frequency of v_7, 950 cm^{-1}, which is the same for the free molecule.

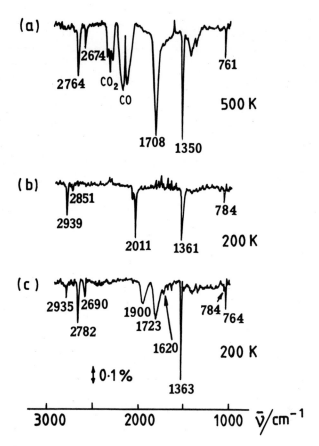

Figure 4 RAIR spectra of formate species on Ru(001); reproduced from reference (14) with permission.
(a) Reaction intermediate observed during CO/H_2 reaction at high pressure over K-Ru(001).
(b) Formate from the decomposition of formic acid on clean Ru(001).
(c) Potassium formate from the decomposition of formic acid on K-Ru(001).

The effect of increasing equilibrium pressure, fig.(5f), is to broaden and reduce the integrated intensity of the v_7 band and to produce the sudden appearance of a band at 3075 cm^{-1} which may be assigned to the normally infrared active mode of ethene, v_9, which is polarized parallel to the plane of the molecule and perpendicular to the C-C axis, fig.(6). This result indicates a structural change near the monolayer point in which ethene molecules are tilting by rotation about the C-C axis, presumably to allow a greater packing density.

This rotation produces a component of dynamic dipole from v_9 perpendicular to the surface leading to its appearance in the spectrum, and rotates the dynamic dipole moment of v_7 away from the surface normal thus reducing its intensity in the spectrum. Note that the other normally infrared active CH stretching mode of ethene, v_{11} expected at <2990 cm^{-1}, produces a dynamic dipole moment parallel to the C-C axis and hence would not become allowed as a result of such a structural change.

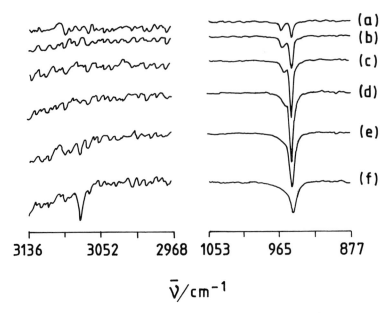

Figure 5 RAIR spectra of ethene adsorbed on Ag(100) at 100K and increasing equilibrium pressure. (a) 1.4×10^{-9} mbar; (b) 1.2×10^{-8} mbar; (c) 1.6×10^{-7} mbar; (d) 2.1×10^{-6} mbar; (e) 1.3×10^{-5} mbar; (f) 5.3×10^{-5} mbar.

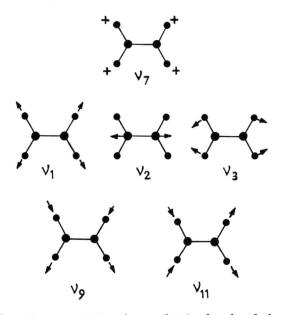

Figure 6 Schematic representation of some vibrational modes of ethene: v_7 is strongly infrared active and allowed by the metal surface selection rule; v_1-v_3 are normally infrared inactive but allowed by the metal surface selection rule by virtue of the ethene-surface interaction. v_9 and v_{11} are normally infrared active but forbidden by the metal surface selection rule if the plane of the ethene molecule is parallel to the surface.

The effect of adsorbed chlorine is illustrated in fig.(7) in which spectra at an equilibrium pressure of 2×10^{-7} mbar are shown for a range of chlorine surface coverages. The ν_7 band is shifted away from the gas phase frequency indicating a stronger perturbation of the ethene molecule. The band at 970 cm^{-1} is stable to pumping, unlike the 950 cm^{-1} band on 'clean' Ag(100), and the associated ethene species desorbs above 135K. The significant degree of enhancement of the silver-ethene interaction illustrated by these results is one possible factor in the chlorine promotion of ethylene oxidation over silver catalysts.

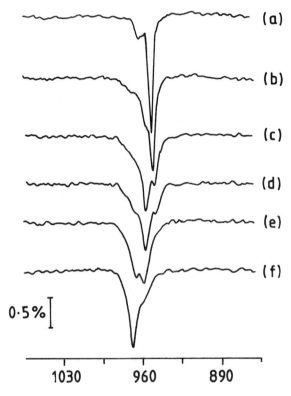

Figure 7 RAIR spectra of ethene adsorbed on Cl/Ag(100) at 100K and 2×10^{-7} mbar equilibrium pressure. Chlorine coverages (a) zero; (b) 0.17; (c) 0.29; (d) 0.40; (e) 0.50; (f) 1.0.

4 RAIRS TOWARDS THE FAR-IR

We have previously referred to the potential importance of the region of the infrared spectrum between 800 and 100 cm^{-1} where one finds vibrational modes associated with adatoms or frustrated translations and rotations of adsorbed molecules. There are two problems associated with working in this region. Firstly the most widely-used detectors in mid-infrared spectroscopy are high sensitivity, narrow band, mercury cadmium telluride detectors with threshold energies at approximately 700 cm^{-1} and to extend this range while retaining sensitivity necessitates use of rather specialized liquid helium cooled photon detectors such as copper/germanium or of liquid helium cooled bolometer detectors. In addition the conventional KBr or CsI beamsplitters, in mid-infrared FT spectrometers, cut off at 400 and 200 cm^{-1} respectively and so ideally a switch to far-infrared optical

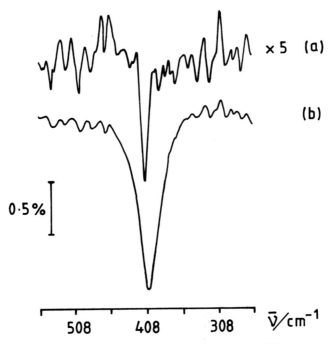

Figure 8 RAIR spectra of ammonia adsorbed on Ag(100) at 100K. (a) Approximately 2 monolayers; (b) multilayer.

hardware is required. Given the resources and the patience these difficulties can be overcome but the second problem is more fundamental. The brightness of a conventional thermal infrared source falls rapidly below 1000 cm^{-1}, e.g. the power available at 2000 cm^{-1} from a 3000K black-body radiator is 30 times that available at 300 cm^{-1}. A possible answer is to use a synchrotron source which has the polarization and low beam divergence required by the RAIRS experiment.[15-17] Comparisons of power levels delivered to a crystal surface in a grazing angle configuration suggest that use of the synchrotron source becomes favourable below approximately 1000 cm^{-1} and the advantage is approximately two orders of magnitude at 300 cm^{-1}.[17]

There are several examples in the literature of successful detection of RAIRS bands at or below 400 cm^{-1} using conventional i.r. sources[18,19] or by infrared emission from a sample at room temperature[20] (using a liquid nitrogen cooled spectrometer!)

A RAIRS experiment has been set up at the Brookhaven Synchrotron and is producing encouraging results.[21] An infrared beam port is under construction at Daresbury and the vacuum and spectroscopic system is currently under test in our laboratories at East Anglia where we have been able to optimize performance using a conventional infrared source and a far infrared spectrometer (the Nicolet 20F). The performance level is illustrated in fig.(8) by the spectra of adsorbed ammonia on Ag(100). The sharp band at approximately 410 cm^{-1} is assigned to a frustrated rotation of a hydrogen bonded ammonia molecule and appears after completion of the first monolayer. It is therefore associated with hydrogen bonding between the first and second adsorbed layers. Further adsorption leads to a shifted and broadened peak associated with solid ammonia. This level of performance is encouraging bearing in mind the expected enhancement from using a synchrotron source.

5 CONCLUSIONS

The RAIRS technique is beginning to fulfil its promise as an *in situ* method of identifying surface intermediates in catalytic processes. Continued improvements in sensitivity will be required to cope with the full range of surface species as infrared intensities depend strongly on chemical structure.

The extension to the 100-800 cm^{-1} will require synchrotron sources except for the very strongest i.r. absorbers in this region.

Recent improvements in the resolution of the EELS technique suggest that the two techniques will continue to enjoy a complementary relationship.

ACKNOWLEDGEMENTS

We would like to acknowledge the support of the Science and Engineering Research Council in equipment grants and a post-doctoral research assistantship (D.A. Slater). We are also indebted to Prof. Norman Sheppard for many useful discussions.

REFERENCES

1. M.A. Chesters, J.Electron Spect., 1986, 38, 123.
2. A.M. Bradshaw and E. Schweizer, 'Spectroscopy of Surfaces', Eds. R.J.H. Clark and R.E. Hester, Adv. in Spectroscopy, 1988, 16, 413.
3. R.G. Greenler, J.Vac.Sci.Technol., 1975, 12, 1410.
4. P.D. Gardner, Ph.D. Thesis, UEA 1988
5. H. Ibach and D.L. Mills, 'Electron Energy Loss Spectroscopy and Surface Vibrations', Academic Press, London, 1982.
6. M.A. Chesters and N. Sheppard, 'Spectroscopy of Surfaces', Eds. R.J.H. Clark and R.E. Hester, Adv. in Spectroscopy, 1988,16, 377.
7. J.E. Demuth, H. Ibach and S. Lehwald, Phys.Rev.Lett., 1978, 40, 1044.
8. M.A. Chesters and P. Gardner, Spectrochim.Acta, 1990, 46A, 1011.
9. R. Raval, S. Haq, G. Blyholder and D.A. King, J.Electron Spect. and Related Phenom., 1990, 54/55, 629.
10. G. Kisters, J.G. Chen, S. Lehwald and H. Ibach, Surface Sci., 1991, 245, 65.
11. Michael A. Chesters, Carlos de la Cruz, Peter Gardner, Elaine McCash, Paul Pudney, G. Shahid and Norman Sheppard, J.Chem.Soc. Faraday Trans., 1990, 86, 2757.
12. Adrian O. Taylor and J. Pritchard, J.Chem.Soc. Faraday Trans., 1990, 86, 2743.
13. F.M. Hoffmann, M.W. Weisel and C.H.F. Peder, J.Electron Spectrosc. and Related Phenom., 1990, 54/55, 779.
14. M.D. Weisel, J.G. Chen and F.M. Hoffmann, J.Electron Spectrosc. and Related Phenom., 1990, 54/55, 787.
15. J. Yarwood, T. Shuttleworth, J.B. Hasted and T. Nauba, Nature, 1984, 312, 742.
16. E. Schweizer, J. Nagel, W. Braut, E. Lippert and A.M. Bradshaw, Nucl.Instrum. Methods, 1985, A239, 630.

17. D.A. Slater, P. Hollins, M.A. Chesters, J. Pritchard, D.H. Martin, M. Surman, D.A. Shaw and I.H. Munro, Rev.Sci. Instrum., 1992, 63, 1547.
18. D. Hoge, M. Tushaus, E. Schweizer and A.M. Bradshaw, Chem.Phys.Lett., 1988, 151, 230.
19. I.J. Malik and M. Trenary, Surface Sci., 1989, 214, L237.
20. R.G. Tobin and P.L. Richards, Surface Sci., 1987, 179, 387.
21. C.J. Hirschmugl, G.P. Williams, F.M. Hoffmann and Y.S. Chabal, J.Electron.Spectrosc. and Related Phenom., 1990, 54/55, 109.

Thermal Evolution and Hydrogenation of Ethylidyne on Pt(111) Studied by RAIRS

Gordon McDougall and Hilary Yates
DEPARTMENT OF CHEMISTRY, UNIVERSITY OF EDINBURGH,
WEST MAINS ROAD, EDINBURGH EH9 3JJ, UK

SUMMARY

In situ reflection absorption infrared experiments directly yield both activation energies for the formation and subsequent decomposition of ethylidyne species on Pt(111) and rates of hydrogenation of the ethylidyne adlayer as a function of hydrogen pressure. At pressures above approximately 1 mbar there is evidence for the population of an 'on top' form of adsorbed hydrogen, the presence of which brings about the onset of ethylidyne hydrogenation.

1 INTRODUCTION

The formation of a surface ethylidyne species (M_3CCH_3, where M=a surface metal atom) from the adsorption of ethene on Pt(111)[1-4] and on a variety of other metal single crystal surfaces[5-7] and supported metal catalysts[8-13] is now well established. The almost ubiquitous nature of this particular hydrocarbon fragment has led to the suggestion that alkylidynes may have a general role in olefin hydrogenation[14]; however, as yet, few experiments have attempted to explore the chemistry of the alkylidynes under realistic reaction conditions. Here we employ Reflection Absorption Infrared Spectroscopy (RAIRS) for the *in situ* observation of both the hydrogenation of ethylidyne at pressures up to *circa* one atmosphere and the thermal evolution of ethylidyne during temperature programmed surface reaction.

2 EXPERIMENTAL

All the experiments described in this work were performed on an apparatus consisting of an Ultra High Vacuum (UHV) chamber, equipped with an isolable infrared cell, coupled to a Fourier transform infrared (FTIR) spectrometer. The system, described in detail elsewhere[15], has the essential features noted below.

The particular spectrometer employed was a Biorad FTS-40, using a standard, narrow band, Mercury Cadmium Telluride detector. The optical bench of the spectrometer and the detector optics housing were evacuable to 10^{-3} mbar and connected directly to the window flanges of the infrared cell giving a completely evacuable infrared beampath.

The UHV system and infrared cell were constructed by CVT Ltd., Milton Keynes. The cell could be operated either under UHV or at pressures up to approximately two atmospheres when isolated from the main UHV chamber by means of a differentially pumped double 'o' ring seal. The upper operating pressure of the cell was largely dictated by the KBr window material. In addition to the infrared cell, the UHV chamber was also equipped with the normal low energy electron diffraction, Auger electron spectroscopy, mass spectroscopy and Ar+ ion bombardment facilities for sample cleaning and characterization.

The Pt(111) single crystal sample was purchased from Goodfellow Metals Ltd., Cambridge. Initial cleaning of the sample required heating to around 650°C under 5×10^{-6} mbar of O_2 for a minimum of six hours to remove heavy sulphur and carbon contamination. Thereafter cycles of Ar+ ion bombardment (6mA at 2keV) at ambient temperatures and annealing *in vacuo* at 650°C 900-1000K were sufficient to routinely produce a clean Pt(111) surface.

Hydrogen and Argon (UHP grade) were supplied by Union Carbide Speciality Gases. Ethene and Oxygen (CP grade) were obtained from Cambrian Gases, Croydon.

3 RESULTS

Figure 1 shows a series of RAIR spectra recorded during temperature programming of hydrocarbon adsorbed on the Pt(111) sample. Each spectrum is the ratio of a single beam spectrum, composed of 500 coadded individual scans and recorded at the temperature shown, against a background single beam of the clean Pt sample recorded immediately prior to exposure to 5×10^{-9} mbar of ethene for 1000 seconds at -110°C. The spectral intensities are quoted as "% transmittance"; the ratio of the sample reflected intensity, I, against the reference reflected intensity of the sample prior to adsorption, I^o, and are directly proportional to the function $\Delta R/R$ often used to denote band intensities in RAIR spectra[16].

250 seconds of data acquisition were required for each spectrum. The

spectra were recorded isothermally and at 4 cm-1 resolution with the temperature raised between each spectrum giving a stepped heating programme approximating to a linear heating rate of 0.04°Cs-1.

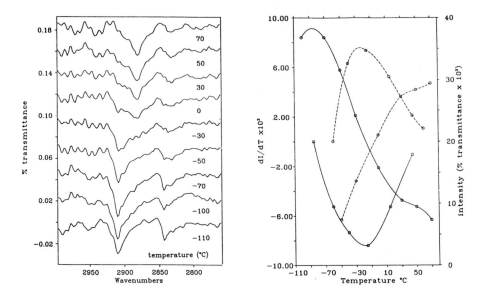

Figure 1 RAIR spectra showing the thermal evolution of ethene adsorbed on Pt(111) at -110°C on heating to 70°C. The accompanying graph gives the variation of the spectral intensity, I, with temperature, T, and the derivative, dI/dT. (solid lines, 2910 cm-1, dashed lines 2880 cm-1)

Two weak features are evident in the low temperature spectra at 2910 and 2840cm-1 which evolve on heating into the single peak at 2880cm-1. The single, higher temperature peak can be assigned to the υCH₃(s) mode of ethylidyne[3,4,17]. At low temperatures ethene is normally considered to adsorb on Pt(111) as an associatively σ-diadsorbed species[2,6,18] and so the twin peaks appearing upon initial adsorption and persisting to almost 283K would be consistent with υCH₂(s) (2910cm-1) and the totally symmetric component of the overtone 2δCH₂(as)(2840cm-1), brought up in intensity by Fermi resonance with the adjacent υCH₂(s), of a surface σ-diadsorded ethylenic species.

The plot accompanying the spectra in Figure 1 shows both the decay of the intensity of the υCH₂(s) mode of the σ-diadsorbed species with increasing temperature and the derivative of the band intensity with respect to temperature, dI/dT (solid lines). The corresponding data for the growth of the υCH₃(s) mode of the ethylidyne species is given by the dashed lines. The T_{min}, or T_{max} as appropriate, in the dI/dT curves give the temperatures where $d^2I/dT^2=0$. Following Redhead[19], and assuming first order kinetics and a frequency factor $A=10^{13}s^{-1}$, the activation energies for the decomposition of the σ-diadsorbed species and formation of the surface ethylidyne fragment are 18±3 and 17±3 kcal mol^{-1} respectively.

The spectra presented in Figure 2 show the $\delta CH_3(s)$ mode of the ethylidyne ligand on the Pt(111) surface and the consequent decay of the band intensity on further heating. The initial spectrum (not shown) was produced by exposing the sample to 5x10^{-9}mbar ethene for 1000 seconds at room temperature to give the saturation coverage 2x2 ethylidyne adlayer[1]. The spectra begin to decay rapidly only on heating to temperatures in excess of *circa* 100°C, with no additional features appearing that might be assigned to decomposition products of ethylidyne. Data acquisition times, spectral resolution and heating programme were as before.

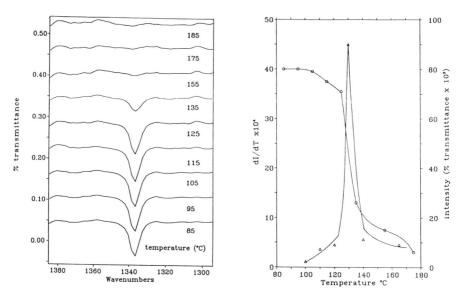

Figure 2 RAIR spectra showing the thermal evolution of ethylidyne on P(111) during heating between 85 and 185°C. The accompanying graph gives the change in the band intensity, I, with increasing temperature, T, and the derivative, dI/dT.

The $\delta CH_3(s)$ mode of ethylidyne was also evident in the experiments discussed in Figure 1 , however, the particular set of spectra was recorded prior to the completion of the evacuable instrument beampath. The near coincidence of a water vapour miscancellation feature, introduced by the spectral ratioing procedure, made accurate peak intensity measurements difficult in these spectra. The data shown in Figure 2 was collected after full commissioning of the instrument and so suffers no such problem. As before, careful measurement of the band intensity with increasing temperature gives T_{min} for the decay of the surface ethylidyne fragment yielding an activation energy for ethylidyne decomposition of 29 ±1 kcal mol^{-1}.

Figure 3 shows the effect of increasing pressures of hydrogen on the appearance of the $\delta CH_3(s)$ mode of the ethylidyne species. Once again the initial spectrum corresponds to saturation coverage of ethylidyne generated by adsorption at room temperature. The individual spectra now represent the coaddition of 1000 scans at 2 cm^{-1} resolution (1000 seconds data acquisition).

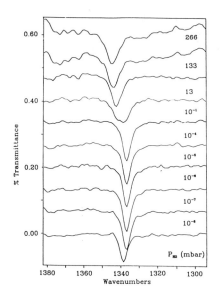
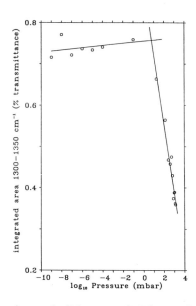

Figure 3 RAIR spectra of $\delta CH_3(s)$ of surface ethylidyne under increasing pressures of H_2. The accompanying graph shows the variation in band intensity with H_2 pressure over the range 1x10^{-8} to 1333 mbar.

Each spectrum shown in Figure 3 was recorded at a constant H_2 pressure with the pressure stepped to the value indicated between each data collection. At low pressures the spectra show little change, other than a *circa* 2cm^{-1} shift in the band position to lower wavenumber and a slight net increase in the peak intensity. However, on exposure to 10^{-1}mbar of H_2, the peak broadens markedly, shifts almost 6cm^{-1} back to higher wavenumber and begins to show a reduction in intensity. The adjacent plot, giving the integrated band intensity under increasing hydrogen pressures up to 1333 mbar H_2, clearly illustrates the same sharp onset of hydrogenation of the surface ethylidyne species.

The rate of hydrogenation was readily observable at H_2 pressures beyond one atmosphere and a typical series of "concentration-time" curves at pressures between 20mbar and 1333mbar H_2 are shown in Figure 4.

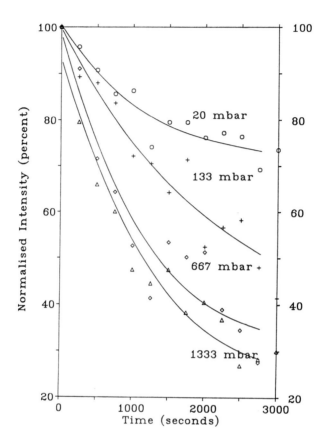

Figure 4 Hydrogenation of surface ethylidyne on Pt(111) under pressures of H_2 between 20 and 1333 mbar. The surface concentration of ethylidyne is expressed as the normalised intensity of the $\delta CH_3(s)$ mode.

Each curve is derived from a set of 4cm^{-1} resolution spectra showing the decay of the concentration of the surface ethylidyne, expressed as the normalised integrated intensity of the $\delta CH_3(s)$ mode, with time. Analysis of this data by standard methods gives initial rates of reaction ranging from 0.9×10^{11} ethylidyne cm^{-2} s^{-1} at 20mbar to 3.0×10^{11} under 1333mbar of H_2 (based on a saturation ethylidyne coverage of 6×10^{14} ethylidyne per cm^2 Pt(111) surface) and indicate an order of reaction with respect to the H_2 pressure of 0.3 ± 0.1. The order with respect to the surface ethylidyne species appears to vary during the course of the reaction, only approaching unity at high conversions.

4 DISCUSSION

Thermal Evolution of Ethylidyne.

A variety of mechanisms have been put forward for the transition between the low temperature, σ-diadsorbed phase of adsorbed ethene on Pt(111) and ethylidyne[20-24]. These include initial dehydrogenation to vinyl (MCHCH$_2$), or vinylidene (M$_2$CCH$_2$) surface fragments, followed by partial rehydrogenation, or isomerization of the σ-diadsorbed species to ethylidene (M$_2$CHCH$_3$) with subsequent elimination of a H atom. The spectra shown in Figure 1 would appear to show a direct transition from σ-diadsorbed ethene to ethylidyne with no evidence for stable intermediate species. This observation does not of course exclude the involvement of vinyl, vinylidene or ethylidene intermediates but merely indicates that the activation energies for the reaction of these intermediates, if formed, must be sufficiently smaller than the 18kcal mol^{-1} required for the initial activation of the σ-diadsorbed species to make their subsequent reaction to ethylidyne facile on the timescale of the RAIRS experiments. Recent theoretical studies place a value of between 18[25] and 21[21] kcal mol^{-1} on E_a for the decomposition of the σ-diadsorbed species, in reasonable accord with the value noted here.

The value of the activation energy for the decomposition of ethylidyne (29 kcal mol^{-1}) is again in good agreement with that estimated by semi-empirical methods (31 kcal mol^{-1})[25] or molecular orbital calculations on a Pt(111) slab (34 kcal mol^{-1})[21]. Electron energy loss vibrational spectroscopy studies[26] have suggested several further surface species to arise from the decomposition of ethylidyne. It is a disappointing outcome of these experiments that no spectral features were observed on further heating beyond the loss of the ethylidyne species.

Ethylidyne Hydrogenation

At pressures less than ~1mbar, exposure to H$_2$ brings about only subtle changes in the intensity and band position of the δCH$_3$(s) mode of ethylidyne. While these changes may indicate a partial hydrogenation to other methyl containing surface fragments such as ethylidene, it seems more likely that the wavenumber shift and intensity increase are due to small changes in the polarization of the surface ethylidyne arising from coadsorption of hydrogen. This hydrogen is most likely to be located at 2 or 3 fold multiply bridged sites[27-29]. Above ~1mbar, however, there is an irreversible drop in the band intensity undoubtedly associated with the loss of ethylidyne via hydrogenation. At these pressures there is considerable evidence from studies of supported catalysts which suggests that hydrogen adsorbs reversibly in an "on top" site on Pt surfaces[30-32] and we believe that it is the appearance of this H species that is responsible for the onset of the hydrogenation of ethylidyne.

Previous evidence on the stability of ethylidyne to hydrogenation comes from two sources, transmission infrared studies of supported metal catalysts by Yates et al.[8] and analysis of single crystal systems before and after ethylene hydrogenation reaction at high pressures[33,34]. The initial rates of ethylidyne hydrogenation found here are in almost exact agreement with those quoted by Yates and directly contradict the apparent stability of ethylidyne under H_2 pressures noted in the "before and after" single crystal work.

Typical orders of reaction during ethene hydrogenation for the hydrogen and ethene reactants are approximately 1 and 0 (or small and negative) respectively for reaction over either supported catalysts[35] or single crystals[33]. The very different values found here for the ethylidyne hydrogenation, together with the very low turnover, confirm the accepted wisdom that ethylidyne itself is not the principal precursor to ethane in the ethene hydrogenation reaction. Since the ethylidyne species would be liable to removal by hydrogenation during the ethene hydrogenation reaction under conditions of excess H_2, it seems reasonable to also conclude, like Yates[8], that ethylidyne is not a necessary intermediary in ethene hydrogenation.

5 REFERENCES

1. L. L. Kesmodel, L. H. Dubois and G. A. Somorjai, J. Chem. Phys., 1979, 70, 2180.
2. H. Steininger, H. Ibach and S.Lehwald, Surf. Sci.,1982,117, 341.
3. M. A. Chesters and E. M. McCash, Surf. Sci., 1987, 187, L639.
4. I. J. Malik, M. E. Brubaker, S. B. Mohsin and M. Trenary, J. Chem. Phys, 1987, 87, 2862.
5. B. E. Koel, B. E. Bent and G. A. Somorjai, Surf. Sci.,1984, 146, 211.
6. L. L. Kesmodel and J. A. Gates, Surf. Sci., 1981, 111, L747.
7. M. A. Barteau, J. Q. Broughton and D. Menzel, Appl. Surf. Sci., 1984.
8. T. P. Beebe, Jr., M. R. Albert and J. T. Yates, Jr., J. Catal., 1985, 96, 1.
9. P. K. Wang, C. P. Slichter and J. H. Sinflet, J. Phys. Chem., 1985, 89, 3606.
10. B. J. Bandy, M. A. Chesters, D. I. James, G. S. McDougall, M. E. Pemble and N. Sheppard, Philos. Trans. R. Soc. London. A., 1986, 318, 141.
11. N. Sheppard, D. I. James, A. Lesiunas and J. D.Prentice, Commun. Dep. Chem. (Bulg. Acad. Sci.), 1984, 17, 95.
12. P. D. Holmes, G. S. McDougall, I. C. Wilcock and K. C. Waugh, Catalysis Today, 1991, 9, 15
13. G. S. McDougall, Ph.D. Thesis, University of East Anglia, 1984.
14. R. J. Koestner, M. A. Van Hove and G. A. Somorjai, J. Phys. Chem., 1983, 87, 203.
15. H. T. Yates, Ph.D. Thesis, The University of Edinburgh, 1992.
16. B. E. Hayden, 'Vibrational Spectroscopy of Molecules on Surfaces' (J. T. Yates, Jr. and T. E. Madey eds.), Plenum Press, New York, 1989.
17. P. Skinner, M. W. Howard, I. A. Oxton, S. F. A. Kettle, D. B. Powell and N. Sheppard, J. Chem. Soc. Faraday. Trans., 1981, 277, 1203.
18. H. Ibach and S. Lehwald, J. Vac. Sci. Technol., 1978, 15, 407.
19. P. A. Redhead, Trans. Faraday Soc., 1961, 57, 641.

20. J. A. Gates and L. L. Kesmodel, Surf. Sci., 1983, 124, 68.
21. D. B. Kang and A. B. Anderson, Surf. Sci., 1985, 155, 639.
22. G. A. Somorjai, M. A. Van Hove and B. E. Bent, J. Phys. Chem., 1988, 92, 973.
23. R.G. Windham and B. E. Koel, J. Phys. Chem. IN PRESS.
24. F. Zaera, J. Am. Chem. Soc. IN PRESS.
25. E. A. Carter and B. E. Koel, Surf. Sci., 1990, 226, 339.
26. A. M. Baro and H. Ibach, J. Phys. Chem., 1981, 74, 4194.
27. A. M. Baro, H. Ibach and H. D. Bruchmann, Surf. Sci., 1979, 88, 384.
28. P. J. Fiebelmann and D. R. Hamann, Surf. Sci., 1987, 182, 411.
29. J. E. Reutt, Y. J. Chabal and S. B. Christman, J. Elec. Spec, 1987, 44, 325.
30. M. Primet, J. M. Basset, M. V. Mathieu and M. Pettre, J. Catal., 1973, 28, 368.
31. T. Szilagyi, J. Catal., 1990, 121, 223
32. M. A. Chesters, A. Dolan, D. Lennon, D. J. Williamson and K. J. Packer, J. Chem. Soc. Faraday Trans., 1990, 86, 3491.
33. F. Zaera and G. A. Somorjai, J. Am. Chem. Soc., 1984, 106, 2288.
34. A. Wieckowski, S. D. Rosasco, G. N. Salatia, A. Hubbard, B. E. Bent, F. Zaera, D. Godbey and G. A. Somorjai, J. Am. Chem. Soc., 1985, 107, 5910.
35. S. J. Thompson and G.Webb, 'Heterogeneous Catalysis', University Chemical Texts, Edinburgh, 1968.

The Adsorption and Reaction of Formic Acid on Oxygen-doped Silver Films: an Infrared Study

S. Munro and R. Raval
DEPARTMENT OF CHEMISTRY, UNIVERSITY OF ABERDEEN,
MESTON WALK, ABERDEEN AB9 2UE, UK

ABSTRACT

The adsorption and reaction of formic acid on oxygen-doped polycrystalline silver films at 170 K has been studied by infrared transmission spectroscopy. Formic acid adsorbs molecularly on clean silver. Scission of the O-H bond to form the formate species occurs only when the surface has been doped with oxygen. Initially, at low exposures of oxygen and formic acid, a bridging formate species is formed with both oxygens interacting with the surface. This species, interestingly, possesses a site symmetry of $C_s(2)$ because the inequivalence of the two surface adsorption sites renders the two oxygens different. Further exposure to formic acid results in the co-adsorption of monomeric formic acid on the surface in C_s symmetry. This formic acid can be converted to the formate species by the adsorption of more oxygen, thus demonstrating that a titration of the adsorbed oxygen atoms occurs during this reaction.

1 INTRODUCTION

Silver is an important catalyst for the selective partial oxidation of many organic molecules, including formic acid. However, studies carried out on the reaction of formic acid on low Miller index planes of silver[1] demonstrate that the scission of the O-H bond in the acid to form the formate ion only occurs in the presence of adsorbed oxygen on the surface. The question that remains is whether such activation of the silver surface is also necessary for polycrystalline silver films where, presumably, a number of higher index planes are also present. We have studied[2] the adsorption of formic acid on clean, vacuum deposited silver films and have found that, as for the low index single crystal planes, only the non-dissociative adsorption of the formic acid takes place on these surfaces.

In this study we have investigated the adsorption and reaction of formic acid on oxygen-doped silver films in order to establish whether the polycrystalline samples also mimic the reactivity behaviour of the oxygen covered single crystal surfaces and form adsorbed formate species. It is known[3] that silver adsorbs dioxygen and partially dissociates it to form adsorbed atomic oxygen. At the temperatures employed in this work, between 50 to 80% of the molecular species should have dissociated to atomic oxygen[3]. Oxygen adsorbed on silver acts as a strong Brønsted base and, consequently, has a strong tendency to cause OH bond scission in catalytic reactions. The reaction of formic acid on oxygen-doped silver surfaces is, therefore, expected to result in the creation of formate, the conjugate base, as a stable intermediate on the surface:-

$$2HCOOH_{ads} + O_{ads} \longrightarrow H_2O_{ads} + 2COOH_{ads}$$
$$\text{or}$$
$$HCOOH_{ads} + O_{ads} \longrightarrow OH_{ads} + COOH_{ads}$$

2 EXPERIMENTAL

Experiments were conducted in a newly constructed ultra-high vacuum (UHV) apparatus, consisting of a stainless steel chamber and a pyrex glass infrared cell. The infrared glass cell is fitted with demountable infrared windows and is capable of withstanding UHV and bakeout conditions[4]. The cell can be positioned in the sample compartment of a commercial Fourier transform infrared spectrometer, allowing the infrared beam to be transmitted through the thin silver film. Prior to evaporation, the source bead is degassed thoroughly, enabling evaporation of the film to occur with a pressure rise of no more than 3×10^{-10} torr. The metal film is evaporated onto a cleaved NaCl plate, cooled to 170 K. Evaporation is stopped when IR transmission through the sample is reduced by 30%.

All spectra are presented as ratios of single beam spectra with the adsorbate present and with the surface clean. The spectra presented here were recorded on the Mattson 6021 FTIR spectrometer at 2 cm^{-1} resolution with 200 scans co-added (accumulation time ~ 3 mins.). A narrow-band mercury cadmium telluride detector was used, allowing the spectral range of 4000-670 cm^{-1} to be accessed.

Formic acid was purified by vacuum distillation and freeze-pump-thaw degassing cycles. Oxygen gas (Masonlite research grade 99.99%) was used without further purification.

3 RESULTS AND DISCUSSION

Figure 1 displays the infrared spectra obtained with increasing coverage of formic acid on a silver film doped with a 10 L exposure of oxygen at 170 K. A comparision of the bands shown in Figure 1a with those of gas-phase and condensed-phase

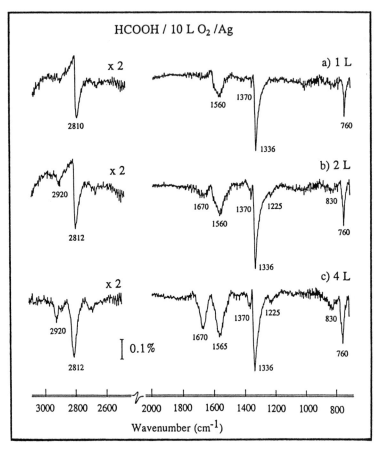

Figure 1. Infrared spectra observed with increasing exposure of formic acid on a silver film doped with 10 L of oxygen. a) 1 L HCOOH; b) 2 L HCOOH; c) 4 L HCOOH.

formic acid[5,6] clearly reveals that the spectrum cannot be attributed to adsorbed formic acid. However, a close correspondence is found with the vibrational spectra obtained for the formate species (Table 1) clearly indicating that this species is created on the oxygen-doped surface.

In order to interpret the infrared results fully, it is necessary to consider the normal modes of vibration of the formate species in more detail. An adsorbed molecule possesses 3N normal modes of vibration since the rotations and translations of the gas phase molecule become hindered rotations and translations of the adsorbed molecule. Therefore, a total of 12 normal modes of vibration exist for the adsorbed formate species. However, the data obtained in this work is restricted to the spectral range of 4000- 670 cm^{-1}: therefore only six normal modes of vibration are pertinent to the analysis. These are illustrated in Figure 2.

Table 1. A comparision of formate vibrational frequencies (cm^{-1}).

Normal mode	[HCOO]$^-$ of sodium salt[8]	formate on Ag{110}[1]	formate on Ni{110}[9]	formate on Cu{100}[10]	This work
ν(C-H)	2841	2900	2864	2840, 2910	2810
ν_a(COO)	1567	1640	1544	1640*	1560
ν_s(COO)	1366	1340	1336	1330	1336
δ(C-H)	1377	-	1360	-	1370
π(C-H)	1073	1050	1024	-	-
δ(OCO)	772	770	768	760	760

* This band has been otherwise assigned as the ν(C=O) vibration of co-adsorbed formic acid[17].

For molecules adsorbed on the silver films, the vibrations observed will be governed by the metal-surface selection rule[7,18] which states that a normal mode of vibration will be infrared active only if the dipole moment change possesses a component perpendicular to the surface. In terms of group theory, this effectively means that only those vibrations that transform as the totally symmetric representation will be allowed. Therefore, the number of allowed vibrations will be determined by the symmetry or point group of the adsorbed molecule which is governed by both the site of adsorption and the orientation of the molecule in that site. The highest symmetry possible for the formate species on the surface is C$_{2v}$ which occurs when the formate is adsorbed in an orientation where the C-H axis and the OCO plane are normal to the surface and the two oxygen atoms are symmetrically equivalent, Figure 3. C$_{2v}$ symmetry may be lowered to C$_s$ by tilting the molecule in one of two directions. These may be differentiated in terms of a tilt of the OCO plane towards the surface giving C$_s$(1) symmetry or a tilt of the C-H axis towards the surface giving C$_s$(2) symmetry, Figure 3. Tilting in both directions simultaneously results in a further lowering of symmetry to C$_1$.

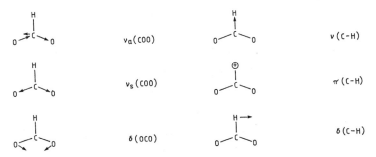

Figure 2. The normal modes of vibration of formate.

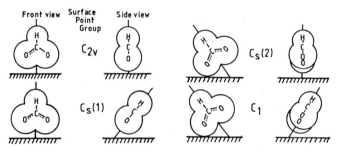

Figure 3. Possible adsorption symmetries that can be adopted by the formate species.

The orientation of the formate species on the oxygen-doped silver films can now be determined by comparing the infrared spectra obtained with the predicted infrared activities in Table 2. The bands in Figure 1a, observed after an exposure of 1L of formic acid, can be assigned as 2810 cm^{-1} ν(C-H); 1560 cm^{-1} ν_a(COO); 1336 cm^{-1} ν_s(COO); 1370 cm^{-1} δ(C-H) and 760 cm^{-1} δ(COO). It should be noted that the derivative shape exhibited by some of the bands has been observed for infrared spectra of other adsorbates on thin films[11].

The presence of the ν_a(COO) band immediately indicates that the symmetry of the adsorbed formate is no higher than $C_s(2)$. This is further confirmed by the presence of the δ(C-H) band which, though weak in intensity, is observed consistently. As shown in Figure 3, this suggests that the formate is adsorbed in the monodentate form with the C-H axis tilted towards the surface and the two oxygens no longer equivalent. However, the nature of formate bonding is also revealed by the frequency difference between the ν_a(COO) and the ν_s(COO) modes[12] as shown in Figure 4. In this case this difference is ~ 224 cm^{-1} and suggests that the formate is bonded in the bridging form with no significant tilt of the C-H axis. In order to reconcile these seemingly conflicting conclusions, we propose that the lowering of

Table 2. Correlation table for the normal modes of vibration of formate under differing adsorbate symmetry (dipole active modes are underlined).

Normal mode	Vibrational band (cm^{-1}) sodium salt[8]	Adsorbate Symmetry			
		C_{2v}	$C_s(1)$	$C_s(2)$	C_1
ν(C-H)	2841	$\underline{A_1}$	$\underline{A'}$	$\underline{A'}$	$\underline{A}$
ν_a(COO)	1567	B_1	A''	$\underline{A'}$	$\underline{A}$
ν_s(COO)	1366	$\underline{A_1}$	$\underline{A'}$	$\underline{A'}$	$\underline{A}$
δ(C-H)	1377	B_1	A''	$\underline{A'}$	$\underline{A}$
π(C-H)	1073	B_2	$\underline{A'}$	A''	$\underline{A}$
δ(OCO)	772	$\underline{A_1}$	$\underline{A'}$	$\underline{A'}$	$\underline{A}$

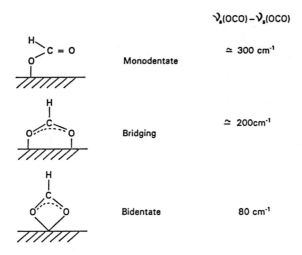

Figure 4. The effect of formate bonding geometry on [v_a(COO) -v_s(COO)].

symmetry to $C_S(2)$ arises, not because of C-H bond tilting, but as a consequence of the inequivalence of the two *surface adsorption sites* in the bridging formate complex which effectively renders the two oxygens different. The presence of inequivalent surface sites is not unexpected for a polycrystalline silver film with a low coverage of adsorbed oxygen, probably present as a disordered lattice gas.

As the exposure of formic acid is increased, new bands appear in the infrared spectra, Figure 1b,c. The small band at 830 cm^{-1} may be assigned[13] to the librational mode of adsorbed OH or H_2O, produced during the creation of formate. The absence[14] of the H_2O scissoring vibration favours assigning this band to a bent adsorbed hydroxyl species. Although, this band is too weak to be observed in Figure 1a it is present in significant concentration when sufficient formate has been generated. The other new bands in Figures 1b, c can be attributed to the adsorption of monomeric formic acid on the surface. We note that the dosing conditions used in this experiment would generate a significant proportion of dimeric formic acid in the gas exposed to the surface[15]. However, we rule out the possibility that the formic acid is adsorbed as a dimer on the grounds that there is no evidence of the broad v(O-H) envelope extending from 3500 - 2500 cm^{-1} which fingerprints the dimer species[5]. The behaviour of the band at 2920 cm^{-1} suggests that it contains contributions from both monomeric formic acid (v(C-H)) and the formate species (various combination bands[16]).

The assignments of the other bands are shown in Table 3. The absence of the π(C-H) band at 1033 cm^{-1} suggests that the symmetry of the adsorbed monomer

Table 3. Correlation table showing how the normal modes of monomer formic acid transform under differing adsorbate symmetry. (The symmetry species of the dipole active modes have been underlined).

Normal mode	Vibrational frequency[5] (cm-1)	Vibrational frequency (this work)	Adsorbate Symmetry C_s	C_1
ν(O-H)	3570*	-	A'	A
ν(C-H)	2943	2920	A'	A
ν(C=O)	1770	1670	A'	A
δ(C-H)	1387	1376	A'	A
δ(O-H)	1229	1225	A'	A
ν(C-O)	1105	-	A'	A
δ(O-C=O)	636*	-	A'	A
π(C-H)	1033	-	A"	A
π(O-H)	636*	-	A"	A

* bands not accessible in this work due to detector characteristics.

is consistent with the C_s point group, Table 3. This implies the molecule is adsorbed with its molecular plane perpendicular to the surface. In addition, the absence of the ν(C-O) vibration at 1105 cm^{-1}, which is a strong gas-phase vibration and allowed in C_s geometry, suggests that the molecule must be adsorbed in the manner depicted in Figure 5, with the C-O bond almost parallel to the surface. Such an adsorption geometry rationalises both the strong intensity of the ν(C=O) band at 1670 cm^{-1} and the weakness of the ν(C-O), ν(C-H) and the δ(O-H) vibrations. In addition, it also explains why the ν(C=O) frequency suffers the greatest perturbation upon adsorption.

We propose that the adsorption of molecular formic acid occurs at this stage because here is a direct titration of adsorbed oxygen during the deprotonation reaction to produce formate. Once the supply of adsorbed oxygen is exhausted, further reaction is halted and non-dissociative adsorption of formic acid occurs.

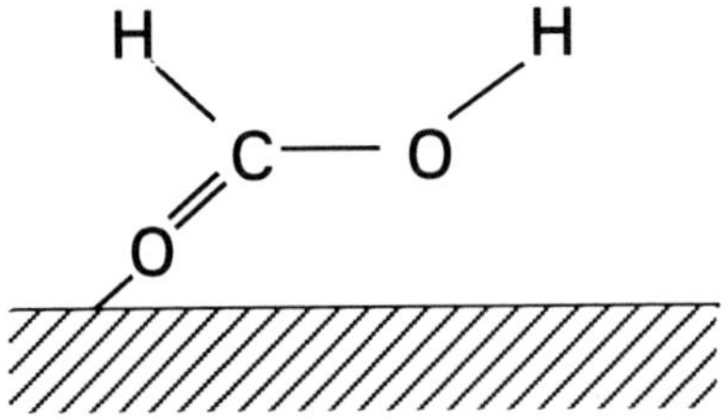

Figure 5. Proposed adsorption geometry of monomeric formic acid.

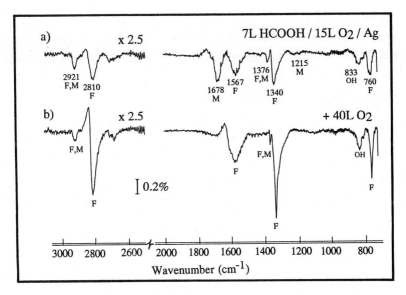

Figure 6. Infrared spectra showing the conversion of monomeric formic acid to formate with increasing exposure of oxygen. F denotes formate bands, M denotes formic acid bands.

This proposition was tested by exposing an oxygen-doped surface to sufficient formic acid so that a significant concentration of both adsorbed formate and monomeric formic acid was created on the surface, Figure 6a. When this system was exposed to further oxygen, a gradual conversion of the formic acid to the formate was observed, until at sufficiently high exposures of oxygen all the formic acid reacted to leave only adsorbed formate on the surface, Figure 6b.

4   CONCLUSIONS

The adsorption and reaction of formic acid on oxygen-doped polycrystalline silver films at 170 K has been studied by infrared transmission spectroscopy. At low exposures of oxygen and formic acid, a bridging formate species is formed with both oxygens interacting with the surface. This species possesses a site symmetry of $C_s(2)$ because the inequivalence of the two surface adsorption sites renders the two oxygens different. Further exposure to formic acid results in the co-adsorption of monomeric formic acid on the surface in C_s symmetry. This formic acid can be converted to the formate species by the adsorption of more oxygen, thus demonstrating that titration of the adsorbed oxygen atoms occurs during this reaction.

Acknowledgments

We are grateful to the SERC, M.O.D and the University of Aberdeen for equipment research grants. It is a pleasure to acknowledge George Beange and Paula Craib for their expert technical assistance throughout this project and Dr. Alison Coats for her help during the initial stages of this project.

REFERENCES

1. B.A. Sexton and R.J. Madix, Surface Sci., 1981, 105, 177.
2. S.Munro and R.Raval, to be published.
3. M.K. Rajumon, K.Prabhakaran and C.N.R. Rao, Surface Sci., 1990, 233, L237.
4. R.Raval, S.Munro, A. Coats and G.Beange, J. Vac. Sci. Technol. A., submitted.
5. R.C. Millikan and K.S. Pitzer, J. Am. Chem. Soc., 1958, 80, 3515.
6. Y-T. Chang, Y. Yamaguchi, W.H. Miller and H.F. Schaefer III, J. Am. Chem. Soc., 1987, 109, 7245.
7. H.A. Pearce and N. Sheppard, Surface Sci., 1976, 59, 205.
8. M. Ito and H.J Bernstein, Can. J. Chem., 1956, 34, 170.
9. T.S. Jones and N.V. Richardson, Phys. Rev. Lett., 1988, 61, 1752.
10. B. Sexton, Surface Sci., 1979, 88, 319.
11. R. Raval and M.A.Chesters, unpublished results.
12. K. Nakamoto, 'Infrared and Raman Spectra of Inorganic and Coordination Compounds', 3rd.Edition, Wiley, New York, 1978.
13. A.F.Carley, P.R. Davies, M.W. Roberts and K.K. Thomas, Surface Sci., 1990, 238, L467.
14. Our recent experiments with DCOOD confirm that there is no contribution from adsorbed H_2O in the 1600 cm^{-1} region.
15. J.B. Benziger and G.R. Schoofs, J. Phys. Chem., 1984, 88, 4439.
16. G. Hussain and N. Sheppard, Spectrochim. Acta, 1987, 43A, 1631.
17. B.E. Hayden, K.Prince, D.P. Woodruff and A.M. Bradshaw, 1983, 133, 589.
18. The metal-surface selection rule has been observed to operate for adsorbates on metal particles of diameter > 20 Å. The data reported in this paper shows the absence of the π(C-H) vibration in both the adsorbed formate and the formic acid species, along with the attenuation of the strong gas-phase formic acid ν(C-O) vibration. This demonstrates that the selection rule also holds for species adsorbed on the evaporated films studied here.

In Situ Characterization of Rhodium Hydroformylation Catalysts

L. Sordelli, R. Psaro, C. Dossi, and A. Fusi
DIPARTIMENTO DI CHIMICA INORGANICA E METALLORGANICA E CENTRO CNR, UNIVERSITÀ DEGLI STUDII, VIA VENEZIAN 21, 20133 MILANO, ITALY

1 INTRODUCTION

The importance of applying *in situ* techniques to the catalyst study needn't be emphasized, since both the surface adsorption mechanisms and the bulk structures of many catalysts depend crucially on the reaction conditions (gas composition, temperature, gas pressure, etc..). Thus, in order to estabilish relationships between the nature of the active centers and the catalytic activity, it's necessary to employ techniques that allow a study of the catalyst surface while the catalytic reaction takes place.
Diffuse Reflectance FT-IR Spectroscopy (DRIFTS) permits observation of the catalyst directly in granular form, overcoming many of the problems associated with the pressed-disc approach [1].
Coupling properly the DRIFT cell with an on-line gas chromatograph allows us to control the catalytic behaviour in the same way as a usual fixed-bed reactor: the DRIFT cell can be regarded as a window into the reactor, providing a view of the catalyst surface under process conditions [2].
An application of such an apparatus is presented here on the spectroscopic characterization of silica-supported rhodium carbonyl clusters, as heterogeneous hydroformylation catalysts.
This reaction has been chosen as model reaction for studying the promotion effect of alkali metal ions on supported rhodium catalysts [3]. We have discovered an active catalyst for ethylene hydroformylation at 453 K and atmospheric pressure when the silica-supported $[Rh_{12}(CO)_{30}]^{2-}$ cluster anion, as K^+ salts, has been used as precursor [4].
The aim of the present study was to investigate, via *in situ* DRIFT spectroscopy, the behaviour of silica-supported $Na_2[Rh_{12}(CO)_{30}]$ under realistic catalytic conditions.

2 EXPERIMENTAL

The $Na_2[Rh_{12}(CO)_{30}]$ cluster has been prepared by the literature method [5]. The Silica (Davison 62, 120 mesh) was impregnated under inert atmosphere from an acetone solution of the cluster, followed by drying in vacuo at room temperature. The air-sensitive solid was stored under inert atmosphere. The metal loading was 1% by weight.

Ethylene hydroformylation was investigated at 453 K and atmospheric pressure ($C_2H_4:CO:H_2$ = 1:1:1) in a continuous-flow glass microreactor interfaced to a HP5890 gas-chromatograph.

Transmission infrared spectra of the catalysts, as pressed wafer, were obtained on a Digilab FTS-40 spectrophotometer. The glass cell used to obtain *in situ* infrared spectra has been described elsewhere [6].

The setup for combined *in situ* DRIFTS and on-line catalytic activity measurement is made up of three main components: i) gas feed/vacuum line system; ii) DRIFTS cell; iii) GC device (HP 5890A) for the analysis of products.

The DRIFTS cell, for controlled temperature and pressure treatments, consists of a modified Harrick HVP vacuum chamber (ZnSe windows) in which the gas ports have been equipped with two valves to make possible charging the cell with the sample in a dry-box (N_2 atmosphere) and then taking it into the spectrometer. The powdered sample, about 50 mg, is placed in the sample holder and the temperature controlled by an Eurotherm DTC 808, monitored with a thermocouple located at its rear. For the measurements the cell is placed in a Harrick DRA-2CI diffuse reflectance attachment ("Praying Mantis") and mounted into the sample compartment of the Digilab FTS-40 spectrometer (DTGS detector, KBr beamsplitter).

DRIFT measurements were carried out at atmospheric pressure in a reaction mixture flow of 10 cc/min, from room temperature to 453 K; the catalytic activity was continuously monitored by the coupled GC.

The spectra were recorded at 4 cm^{-1} resolution, in steady-state flow condition, at increasing temperatures, with steps of typically 10°C, after allowing the system to stabilize for 30 minutes. The coaddition of 100 scans per spectrum gives a satisfactory signal-to-noise ratio, since the beam losses due to the mirrors and windows lower the energy throughput to 5% of the original value. The raw spectra were converted into Kubelka-Munk reflectance form using the spectrum of KBr as reference, and then subtracting the oxide support spectra taken at the same temperature as the sample's ones.

3 RESULTS AND DISCUSSION

The results of ethylene hydroformylation are summarized in Table 1. At 453 K the simple hydrogenation to ethane is predominant over hydroformylation, while the formation of 1-propanol is negligible.
However, the activity for hydroformylation is greatly increased on the catalyst derived from $Na_2[Rh_{12}(CO)_{30}]$. A selectivity of about 30% to aldehyde is observed without any induction period. In addition activity and selectivity for hydroformylation were nearly constant for about 100 h on stream. These results seem to suggest that the molecular anionic cluster is stable under reaction conditions as the catalytically active species. The driving force towards the formation of aldehyde is the intimate contact between the alkaline metal ion and the rhodium framework or, in other words, the presence of the metal promoter in the coordination sphere of the anionic rhodium cluster.
A similar explanation, based on the close contact between rhodium particles and promoter, was proposed by Sachtler [7] to explain the high selectivity to aldehyde of zinc promoted rhodium catalysts.

The isolated Rh^I sites, prepared from the mononuclear $Rh(CO)_2(acac)$, show negligible catalytic activity. This behaviour is in marked contrast with the very high activity of soluble Rh^I complexes for the homogeneous hydroformylation of olefins [8].

An unpromoted catalyst was prepared from $Rh_4(CO)_{12}$; it showed higher initial activity toward hydrogenation, but the selectivity to aldehyde was 20% only. After 50 h the total activity was reduced by 40%, as a consequence of the decrease of hydrogenation activity. Consequently, the two catalytically competitive processes, hydrogenation vs. hydroformylation, require a different topology of supported rhodium.

The very low activity of supported metallic rhodium, prepared in a conventional way, could be explained in terms of rhodium dispersion. XRD studies indicate the presence of metallic particles of about 40 A. A remarkable effect of metal dispersion on hydroformylation over Rh/SiO_2 catalyst has been previously observed [9].

Our catalytic results indicate that neither Rh^I isolated sites or metallic particles of 40 A are the active species for olefin hydroformylation. Recently Hoost et al.[10] reported that, from a statistical treatment of the poisoning phenomenon, the required reaction ensemble for the structure-sensitive ethane hydrogenolysis over potassium-promoted silica supported ruthenium catalysts, was estimated to be made up of about 12 contiguous metal atoms.

TABLE 1 Catalytic activity in ethylene hydroformylation of silica-supported Rh catalysts.

Precursor	Coversion [1] (mol%)		Selectivity to C_2H_5CHO	
	1 h	50 h	1 h	50 h
$Na_2[Rh_{12}(CO)_{30}]$	13.8	11.3	30.7	31.5
$Rh_4(CO)_{12}$	17.7	10.0	20.5	24.8
$Rh(CO)_2 acac$	0.7	0.6	----	----
$RhCl_3$ [2]	1.9	1.7	----	----

Reaction conditions: 1 atm (C_2H_4:CO:H_2 = 1:1:1), 453 K
1. Conversion as (mol ethylene conv.)/(mol ethylene)*100
2. Calcined in O_2 at 673 K, followed by the reduction in H_2 flow at 673 K

Spectroscopic characterization.

When the clusters of $Na_2[Rh_{12}(CO)_{30}]$, as an acetone solution, were adsorbed on a pressed wafer of untreated silica, the initial red violet color of the clusters is mantained. The IR spectrum, in the ν(CO) region, shows relatively broad bands at 2057 and 1840 cm^{-1} (Fig. 1-a), which resemble those of the pure cluster in THF solution [5]. This result indicates that the cluster is physisorbed intact on the silica surface. The broadening of the band due to bridging CO could be related to the occurrence of some weak interactions with the OH groups of the surface, likely via hydrogen bonding.

However, the presence of shoulders at 2089 and 2025 cm^{-1}, characteristic for the isolated Rh^I gem-dicarbonyl sites [11], suggests the oxidative disruption of the rhodium cluster framework by OH groups of the hydrated silica surface. This process slowly occurs in Argon atmosphere at room temperature, being quite complete after 12 h (Fig. 1-b). Exposure to CO at 373 K only increases the intensity of the two bands (Fig. 1-c) at 2089 and 2025 cm^{-1}, without significant modifications of the IR spectrum. In agreement with the known rhodium surface chemistry [11], the $Rh^I(CO)_2$ species can be easily converted to the metallic state by treatment in hydrogen at room temperature.

From this infrared investigation, on pressed wafer, we conclude that the physisorbed anionic rhodium cluster slowly transforms to $Rh^I(CO)_2$ surface species.

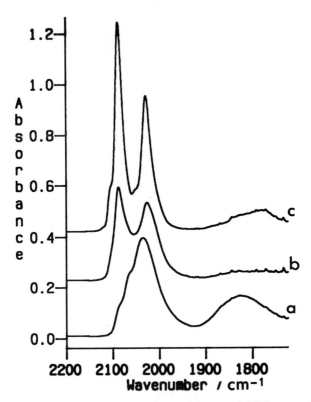

FIGURE 1: IR spectra in the $\nu(CO)$ region of $Na_2[Rh_{12}(CO)_{30}]/SiO_2$: a) physisorbed state after outgassing at 298 K; b) after 12 h in Ar atmosphere at r.t.; c) after 2 h in CO ($5*10^4$ Pa) at 373 K.

Consequently a partial transformation of the supported rhodium cluster into surface Rh^I carbonyl is always taking place during loading of the fresh catalyst into the reactor, although air contamination is avoided (see Experimental). In fact, the DRIFT spectrum recorded at room temperature on stream of reaction mixture ($C_2H_4:CO:H_2 = 1:1:1$) shows the doublet at 2086 and 2030 cm^{-1} (Fig. 2-a). After 1 h at 363 K on stream (Fig. 2-b), almost no appreciable changes are observed with respect to the bands at 2086 and 2030 cm^{-1}, while new bands at 2056 and 1830 cm^{-1} appeared. The new bands are characteristic for CO adsorbed on metallic rhodium particles [11], and very similar to those observed for physisorbed $Na_2[Rh_{12}(CO)_{30}]$.

These results indicate a partial reconstruction of the cluster framework under reaction mixture. This surface behaviour is not unexpected: $Rh_6(CO)_{16}/SiO_2$ is regenerated from the isolated $Rh^I(CO)_2$ sites by treatment with $CO+H_2O$ at room temperature and atmospheric pressure [12]. Above 423 K CO adsorption leads to the formation of rhodium crystallites at the expense of isolated Rh^I sites [13]. In addition, the presence of hydrogen enhances [14] the agglomerizing effect of CO exhibited at and above 423 K.

Since the DRIFT cell has been coupled to on-line GC, it was possible to monitor that the formation of propanal, in the gas phase, starts at 383 K. At the same time, the band due to the linearly adsorbed CO decreases and shifts from 2056 to 2040 cm^{-1}, the broader maximum of the bridged form is around 1800 cm^{-1} (Fig. 2-c). At 453 K, temperature at which the catalytic test is carried out, only a weak broad band around 2035 cm^{-1} is still present (Fig. 2-d). This is the DRIFT spectrum of the surface species under the actual catalytic conditions of the catalytic reaction.

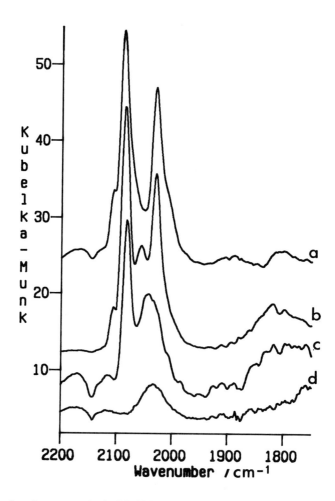

FIGURE 2: Sequential DRIFT spectra of $Na_2[Rh_{12}(CO)_{30}]$ powder in reaction mixture flow: a) at 303 K for 5 min; b) at 363 K for 1 h; c) at 383 K for 1 h; d) at 453 K for 1 h.

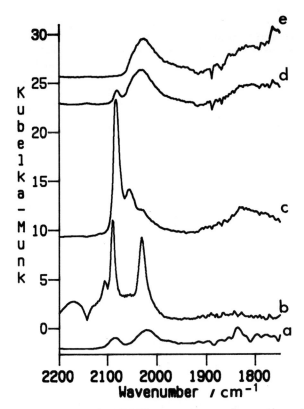

FIGURE 3: Sequential DRIFT spectra of used catalyst: a) after air exposure for 10 min; b) after 10 min in CO flow at r.t.; c) after 30 min in reaction mixture at 373 K; d) as above at 413 K; e) as above at 453 K.

We cannot exclude that sintering to large rhodium crystallites occurs during the catalytic process. The fact that our XRD studies do not detect the presence of metallic rhodium (about 20 Å sensitivity) on used catalyst indicates that Na^+ ions contiguous to rhodium atoms prevent their aggregation.

The results obtained in the present work may indicate an important role of very small metallic particles as active species for olefin hydroformylation or precursors of them. In order to confirm our findings, the hydroformylation reaction was stopped and the catalyst was briefly exposed to air at room temperature. The oxidative fragmentation of the rhodium crystallites leading to isolated $Rh^I(CO)_2$ sites was demonstrated by the appearance of typical doublet at 2090 and 2030 cm^{-1} in the DRIFT spectrum (Fig. 3-b). After exposing the oxidised catalyst to reaction conditions again, the catalytic activity was restored (Table 2), with the DRIFT spectrum (Figs. 3-c,d,e) indicating that the active ensemble of rhodium surface atoms is regenerated.

TABLE 2: The effect of air exposition on the catalytic activity and selectivity of $Na_2[Rh_{12}(CO)_{30}]/SiO_2$ in ethylene hydroformylation.

	Conversion (mol %)		Selectivity to C_2H_5CHO	
	1 h	20 h	1 h	20 h
fresh catalyst	13.8	11.3	30.7	31.5
used catalyst	10.5	9.2	30.0	31.7

4 CONCLUSIONS

We have developed a new DRIFTS cell coupled with a gas chromatograph for *in situ* studies of heterogeneous metal catalysts. The main advantages of this apparatus are: i) operating conditions similar to those in a continuous-flow catalytic reactor; ii) the simultaneous analysis of the gas phase by coupled GC; iii) *in situ* spectroscopic characterization of the catalyst in each of the following steps: activation, reaction and regeneration.

ACKNOWLEDGEMENTS

This work was supported by grants from National Research Council (Project Chimica Fine) and by the Ministry of University and Scientific and Technological Research.

REFERENCES

1. P.D. Holmes, G.S. McDougall, L.C. Wilcock, Catal. Today, 1991, 9, 15.
2. C. Dossi, L. Sordelli, A. Fusi and R. Psaro, in preparation.
3. W.H.M. Sachtler and M. Ichikawa, J. Phys. Chem., 1986, 90, 4752; S. Natio and M. Tanimoto, J. Catal., 1991, 130, 106.
4. C. Dossi, A. Fusi, L. Garlaschelli, D. Roberto, R. Ugo and R. Psaro, Catal. Lett., 1991, 11, 335.
5. P. Chini and S. Martinengo, Inorg. Chimica Acta, 1969, 3, 299.
6. M. Primet, J.M. Basset, E. Garbowski and M.V. Mathieu, J. Am. Chem. Soc., 1975, 97, 3655.
7. M. Ichikawa, A.J. Land, D.F. Shriver and W.H.M. Sachtler, J. Am. Chem. Soc., 1985, 107, 7216.

8. D. Evans, J. Osborne and G. Wilkinson, J. Chem. Soc. A, 1968, 3133.
9. H. Arakawa, N. Takahashi, T. Hanaoka, K. Takeuchi, T. Matsuzaki and Y. Sugi, Chem. Lett., 1988, 1917.
10. T.E. Hoost and J.G. Goodwin, Jr., J. Catal., 1991, 130, 283.
11. J.L. Bilhou, V. Bilhou-Bougnol, W.F. Graydon, G.M. Basset, A.K. Smith, G.M. Zanderighi, R. Ugo, J. Organomet. Chem., 1978, 153, 73.
12. A. Theolier, A.K. Smith, M. Leconte, J.M. Basset, G.M. Zanderighi, R. Psaro and R. Ugo, J. Organomet. Chem., 1980, 191, 415.
13. F. Solymosi and M. Pasztor, J. Phys. Chem., 1985, 89, 4789.
14. F. Solymosi and M. Pasztor, J. Phys. Chem., 1986, 90, 5312.

Pretreatment Effects on CO Oxidation Behavior of Platinum–Rhodium Catalysts – a Combined FTIR, XPS, and TPR Study

James A. Anderson[1,*] and Byron Williams[2]
[1] DEPARTMENT OF CHEMISTRY, THE UNIVERSITY, DUNDEE DD1 4HN, UK
[2] THE LEVERHULME CENTRE FOR INNOVATIVE CATALYSIS, DEPARTMENT OF CHEMISTRY, UNIVERSITY OF LIVERPOOL, LIVERPOOL L69 3BX, UK

Pt-Rh/Al_2O_3 catalysts prepared by single and two stage aqueous impregnation procedures have been compared for the reaction of CO oxidation over a wide range of O_2/CO ratios (0.25-35/1). For catalysts calcined in oxygen prior to hydrogen reduction, no significant difference in catalytic activity was found between the two catalysts despite differences in the nature of the dispersed metals in the two samples as indicated by *in situ* FTIR, XPS and TPR of the precalcined samples. Omission of the calcination stage had negligible effect on the performance of the two stage prepared catalyst but led to a *ca*. 10 K reduction in light-off temperature for the coimpregnated catalyst. Results indicate that sample pretreatments which favor the formation of highly dispersed rhodium phases may produce catalysts which exhibit poorer performance due to particle disruption in the reactant gas mixture and the formation of catalytically inactive dicarbonyl species.

1 INTRODUCTION

Platinum and rhodium are the active components in three-way catalysts currently used for the simultaneous control of nitrogen oxides, carbon monoxide and hydrocarbons from automobile exhaust systems.[1] Many studies of the individual components have been conducted although less is known about the combined behavior of platinum and rhodium in the bimetallic system. Previous studies indicate that single and two-stage impregnations may lead to the formation of Pt-Rh catalysts which display differences in CO oxidation activity and ageing characteristics.[2,3] Of particular interest is the report that two stage impregnation produced a catalyst which displayed a synergistic activity enhancement for CO oxidation relative to a physical mixture of the components and which could not be obtained for the coimpregnated catalyst.[3] Subsequent studies have failed to substantiate such behavior for silica supported Pt-Rh catalysts.[4] The present paper compares alumina supported catalysts prepared by single and two stage aqueous impregnation procedures, and under similar conditions used in examination of the monometallic components[5,6] to determine whether such behavior could be observed and attempts to characterise such catalysts by use of *in situ* FTIR, TPR, and XPS to examine how such behavior may be initiated.

*Present address; Instituto de Catalisis y Petroleoquimica, Campus de la Universidad Autonoma de Madrid, 28049 Madrid, Spain.

2 EXPERIMENTAL

Platinum-rhodium catalyst (0.5wt.% in each metal) prepared by single stage impregnation, [Pt-Rh(sim)], involved aqueous slurries of H_2PtCl_6 and $Rh(NO_3)_3$ with Degussa C γ-Al_2O_3. Pt-Rh catalyst (0.5wt.% in each metal) prepared by two stage impregnation sequence,[Pt-Rh(step)], involved addition of $Rh(NO_3)_3$ solution to a Pt/Al_2O_3 sample which had been precalcined in oxygen at 673 K. Samples were subsequently dried in air at 383 K and stored in vials within an evacuated desiccator. Catalyst samples for infrared spectroscopy were presented in the form of self-supporting discs (25mm diameter, ca. 100 mg) formed by compressing the impregnated alumina between two polished stainless-steel dies (50 MN m^{-2}). Samples were transferred to a vacuum apparatus and heated in a flow of oxygen (60 cm^3 min^{-1}) for 1 h at 673 K before evacuation at the same temperature for 15 min then cooling to ambient temperature under dynamic vacuum. Samples were subsequently transferred to a high temperature stainless-steel infrared cell, which also functioned as a flow reactor, and were evacuated overnight at ca. 300 K. Samples were reduced by heating in a flow (150 cm^3 min^{-1}) of 3.5vol.% hydrogen in argon to 573 K at 5 K min^{-1} and then maintaining this temperature for 2 h followed by evacuation at the high temperature for 15 min, and cooling to 293 K in dynamic vacuum. Reduced catalysts were exposed to the reactant gases in a flow (50 cm^3 min^{-1}, 101 kN m^{-2}) with nitrogen making up the balance and heated in a linear temperature program ramp (10 K min^{-1}) to 573 K. IR spectra (4 cm^{-1} resolution, average of 4 scans) were recorded using a Perkin Elmer 1710 FTIR and 7500 computer. Percentage CO conversions were calculated from the relative intensity of the infrared band at 2361 cm^{-1} due to gaseous CO_2. More detailed information regarding the experimental procedure and apparatus are given elsewhere.[5,6]

Temperature programmed reduction of the two catalysts and also of the two alumina supported monometallic components were performed using 0.3 g of the precalcined (673 K, 1 h) samples. 10vol.% H_2 in nitrogen (30 cm^3 min^{-1}) was passed through the catalyst bed while heating at 10 K min^{-1} from 298-623 K and measuring the thermal conductivity using a Gow-Mac (model 50-350/450) gas analyser. X-Ray photoelectron spectra were measured on a VG ESCA 3 spectrometer with an AlK$_\alpha$ anode ($h\nu$ =1486.6 eV) operating at 400 W. Peak energies were determined relative to the Al 2$s_{1/2}$ peak of the alumina support. Relative surface concentrations were determined by measurement of Al 2$s_{1/2}$, C 1$s_{1/2}$, Cl 2p, Rh 3d$_{5/2}$, O 1$s_{1/2}$ and Pt 4d$_{5/2}$ peak intensities. Following background subtraction[7] the peak areas were integrated numerically and divided by theoretical photoionisation cross-sections[8] and the inelastic mean free path length; the latter calculated from data of Seah and Dench.[9] Spectral deconvolution was performed using a program which allowed mixing of Gaussian-Lorentzian peak shapes and inclusion of tailing information. Following manual fitting, the peak fits were optimised automatically to give the statistical best fit.[10] Rhodium spectra of the Pt-Rh catalyst were fitted using the experimental data for a 1% Rh/Al_2O_3 sample[5] from which a ratio of 1.6-1 was obtained for the Rh 3$d_{5/2}$ and the Rh 3$d_{3/2}$ signals.

3 RESULTS AND DISCUSSION

TPR profiles of the calcined catalysts are shown in fig.1 where the profiles of the bimetallic samples may be compared with those for the monometallic components. Rhodium was reduced in a single stage at 333 K in accordance with the expectation for well-dispersed Rh/Al_2O_3 catalysts,[11] while platinum was reduced in steps at 403 and 448 K possibly indicating reduction of α-PtO_2 and PtO_xCl_y species.[12,13] Reduction of a similar species for Pt-Rh(step) which gave the higher temperature peak for Pt alone may be implicated by the peak at 458 K. However, this peak was absent in the case of the Pt-Rh(sim) catalyst most likely indicating the rhodium catalysed reduction of the platinum complex as previously shown in EXAFS[14]

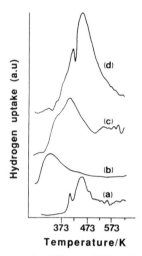

Figure 1. TPR profiles for (a) Pt, (b) Rh, (c) Pt-Rh (sim) and (d) Pt-Rh(step)

and TPR[15] studies of the uncalcined materials. Hydrogen uptake at 413 for Pt-Rh(sim) and 418 K for Pt-Rh(step) are consistent with a single peak at 415 K for $Pt-Rh/Al_2O_3$.[15] Additionally, binding energies of ca. 316 eV for both bimetallic catalysts indicates the presence of Pt^0 in these catalysts following impregnation and 383 K drying and after 673 K oxidation [Table 1]. An apparent shift to higher temperatures for the rhodium reduction peak in the presence of platinum may be attributed to its less facile reduction in the presence of chlorine. Chlorine retention by both catalysts was shown by XPS [Table 2] where the lower Cl:Al ratio for Pt-Rh(step) of 0.012 compared with 0.016 for Pt-Rh(sim) is the result of the additional oxidation treatment of platinum for the former catalyst.[16]

Despite these differences in the nature of the supported phases following oxidation treatment, catalytic activity as shown by the 50% conversion temperatures [fig. 2], of samples which were oxidised and reduced indicate considerable consistency between the two samples over a wide range of $O_2:CO$ ratios. Both samples show a

Table 1. XPS analysis of the platinum and rhodium binding energies (eV) prior to and following calcination at 673 K.

Sample	Rh $3d_{5/2}$	Rh $3d_{3/2}$	Pt $4d_{5/2}$ (metal)	Pt $4d_{5/2}$ (oxide)
Pt-Rh(sim)[a]	308.3	313.7	316.0	318.8
Pt-Rh(sim)[b]	309.7	314.8	315.8	319.3
Pt-Rh(step)[a]	310.0	314.8	315.7	319.6
Pt-Rh(step)[b]	310.1	314.9	316.0	320.0

[a] after impregnation, [b] after calcination at 673 K.

Table 2. XPS analysis of platinum, rhodium and chlorine surface atomic ratios referenced to aluminium for catalysts prior to and following calcination at 673 K.

Sample	Rh	Pt(metal)	Pt(oxide)	Cl
Pt-Rh(sim)[a]	0.0086	0.0026	0.0011	0.027
Pt-Rh(sim)[b]	0.0102	0.0012	0.0015	0.016
Pt-Rh(step)[a]	0.0070	0.0017	0.0012	0.028
Pt-Rh(step)[b]	0.0088	0.0019	0.0016	0.012

[a] after impregnation, [b] after calcination at 673 K.

progressive decrease in 50% conversion temperature with an increase in O_2:CO ratio consistent with expectations of inhibition of the reaction by adsorbed CO under the more CO rich conditions.[6] A slight indication of differences between catalyst behavior was found under the most extreme oxidising conditions used here (35:1), where Pt-Rh(sim) catalyst appeared to suffer the greater detrimental effect in oxygen. As this poorer performance was exhibited over the complete conversion range, it is more probable that this was the result of a structural modification on contact with the high oxygen concentration rather than a deactivation resulting from the formation of oxide which is observed only at conversions of above 50%.[5] As observed in a previous study involving Pt-Rh/Al_2O_3 catalysts,[2] light-off curves for catalysts here lay between those of the monometallic components.

As oxidation treatment below the decomposition temperature of Rh_2O_3 followed by low temperature reduction treatment may result in the formation of separate Pt and Rh phases,[17] experiments were also conducted under similar reaction conditions using catalysts which were reduced without prior oxidation treatment. Results for catalysts in a O_2:CO flow at 4:1 are shown in figure 3 in the form of light-off curves and may be compared with results for samples under similar conditions which had been preoxidised and reduced.

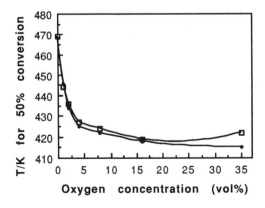

Figure 2. Temperature required to obtain 50% CO conversion as a function of oxygen concentration for Pt-Rh(sim) □ and Pt-Rh(step) ●

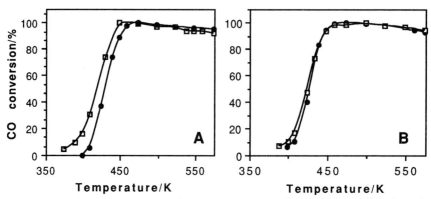

Figure 3. Light-off curves in $O_2(4\%)$-$CO(1\%)$ for (A) Pt-Rh(sim) and (B) Pt-Rh(step) for reduced catalysts with ● and without □ precalcination treatments.

Omission of the oxidation step had negligible effect on the light-off behavior of the Pt-Rh(step) catalyst whereas the light-off curve for Pt-Rh(sim) was shifted to lower temperature resulting in a ca. 10 K reduction in the 50% CO conversion values.

D'Aniello et al.[2] report that for increasing dispersions of Rh/Al_2O_3 and $Pt-Rh/Al_2O_3$ the 50% CO conversion temperature was initally lowered, although at higher dispersions a subsequent increase in light-off temperature occurred. This decreased activity at high dispersion has been attributed to the presence of highly dispersed rhodium phases which result in the formation of inactive dicarbonyl species while in the contact with the reactant gases.[5] Heat treatment of supported rhodium catalysts in CO alone or in net-reducing gas mixtures containing CO, results in an agglomeration of the dispersed phase with a subsequent suppression of the gem-dicarbonyl species and a considerable improvement in activity in the form of reduced light-off temperatures.[5] While such a phenomenum would not be expected during heating of the catalysts in the 4:1 O_2:CO gas mixture used here, pretreatments which subdue the formation of highly dispersed rhodium phases and hence suppress the formation of inactive gem-dicarbonyl species, may be expected to result in catalysts which exhibit relatively improved performance. The ratio of the bands due to rhodium gem-dicarbonyl species to those of the bridged and linear carbonyl species may be used to indicate the relative proportions of highly-dispersed and crystalline metal forms present for each catalyst. Figure 4 shows representative FTIR spectra recorded in situ with the catalyst samples at 423 and 473 K in the gas flow. Bands at ca. 2095 and 2025 cm^{-1} are due to the rhodium gem-dicarbonyl species while the maximum between this pair at ca. 2063 cm^{-1} is assigned to a composite of rhodium and platinum carbonyls linearly bonded to sites on metal crystallites. A band at ca. 2125 cm^{-1} may be assigned to carbonyl species on oxidised forms of platinum and rhodium.[5,6] At the higher of the two temperatures, the linear carbonyl species are eliminated indicating reaction at these sites, while dicarbonyls and carbonyls at oxidised sites remain, indicating the negligible activity of these species.[5,6] From spectra in figure 4, it is clear that dicarbonyl maxima for the

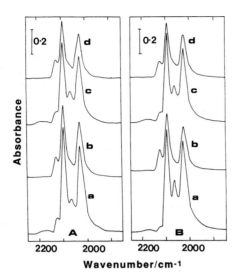

Figure 4. Infrared spectra of catalysts in a flow of $CO(1\%)-O_2(4\%)$ (A) calcined and reduced and (B) reduced only, for Pt-Rh(sim) at (a) 423 and (b) 473 K, and Pt-Rh(step) at (c) 423 and (d) 473 K.

precalcined Pt-Rh(sim) catalyst were relatively more intense than for a similarly pretreated Pt-Rh(step) catalyst, although both samples exhibited similar activities (Fig.2). The same species were subdued for the Pt-Rh(sim) catalyst where the calcination stage was omitted while the linear carbonyl maximum was enhanced (Fig.4). On the other hand, omission of the calcination stage had negligible effect on the proportion of dicarbonyl to linearly bound species formed for Pt-Rh(step) catalyst indicating that the oxidation stage may only lead to enhancement in the proportion of the dispersed rhodium phase for a specific surface configuration of platinum and rhodium species eg. small particles containing rhodium only. One possible explanation for the different effect of precalcination on the two samples would be that formation of bimetallic particles was facilitated for Pt-Rh(sim) by direct reduction of the precursors rather than from the mixture of oxides and reduced metal phases present following oxidation. This would have the effect of reducing the proportion of small particles containing rhodium only. Previous studies indicate that such bimetallic particles may suppress the formation of rhodium gem-dicarbonyl species,[18,19] relative to rhodium only particles although possible differences in particle size should not be ignored.[19] The presence of bimetallic particles would be consistent with analytical electron microscopy (AEM) studies of fresh and aged vehicle exhaust catalysts containing platinum and rhodium.[20,21] Additionally, differences in rhodium detection by AEM and EPMA (electron probe microanalysis) suggested that much rhodium was also present in non-particulate form.[21]

It is unlikely that changes in Pt dispersion are responsible for the observed effects. Stepwise impregnated catalysts performed similarly and gave similar IR spectra whether calcined and reduced (ie. Pt calcined twice) or reduced only (ie. Pt calcined once).

Table 3. Physical characteristics of catalysts following metal salt impregnation and drying but prior to final calcination.

Catalyst	BET area m^2g^{-1}	Micropore area m^2g^{-1}	Micropore volume $(\times 10^3) cm^3g^{-1}$	Average pore diameter nm
Pt-Rh(step)	94.08	3.93	1.74	28.49
Pt-Rh(sim)	86.19	1.47	0.54	30.52

Additionally, previous studies in this laboratory indicate that oxidation at 673 K is ineffective in redispersing Pt on a chlorinated support.[16] Changes in particle morphology had minimal effect on light-off plots for Pt/Al_2O_3 catalysts under equivalent reaction conditions.[6]

As indicated by the relative proportions of gem-dicarbonyl species for the two catalysts, Pt-Rh(step) preparation produced the lower dispersion with respect to the rhodium component in the catalyst. The resultant particles were less influenced by precalcination or treatment in the reactant gas atmosphere than those of the Pt-Rh(sim) catalyst. The reason why larger rhodium particles should be formed by the two stage impregnation than for coimpregnation would appear important in understanding the behavior of the two catalysts. Assuming that during coimpregnation the salts have an equal affinity for the support, then for concentrations used here, monomolecular adsorption should occur.[15] One possibility therefore is that following Pt salt impregnation and calcination, monomolecular adsorption of the rhodium salt cannot take place. This would result in the formation of rhodium salt crystallites and subsequently to the formation of relatively larger rhodium particles following reduction. In a previous study[15] loss in capacity of rhodium salt uptake was attributed to the lowering of pH relative to coimpregnation, the effect of residual chlorine in the support, and the loss of anchorage sites (eg. surface hydroxyl groups) from the support. Loss of the latter may also reduce the tendency for dicarbonyl formation in the presence of CO.[22,23] One possibility, is that following platinum salt impregnation and calcination, partial blocking of the inner pore structure of the alumina occurs leading to resticted access of the support surface for the rhodium salt solution. Alternatively, the additional calcination treatment of Pt-Rh(step) may result in a reduction in BET surface area or affect the crystallinity and pore distribution of the alumina support. A previous study showed that hydrothermal treatment of the support produced a more well-ordered crystalline structure than dry heat treatment.[25] The former support favored the formation of larger Pt-Rh particles and consequently led to lower light-off temperatures for CO oxidation.[25] To investigate this possible support effect, pore size distribution and surface areas of the two catalysts were measured following metal salt impregnation. As shown in Table 3, and in contrast to the above expectations of structural alteration to the alumina as a result of calcination prior to rhodium impregnation, Pt-Rh(step) catalyst shows a relatively higher BET area and micropore area/volume. The latter may be the result of chlorination of the alumina and etching of the structure during heat treatment in oxygen. Therefore any differences in external surface rhodium concentrations for the two catalysts can be considered the result of a chemical rather than a physical effect. In this respect, the influence of chlorine concentration [15] can be discounted as this was higher in the case of Pt-Rh(sim) than for Pt-Rh(step) (Cl:Al,0.016 and 0.012 after calcination). Calculation of the relative proportions of rhodium in the external surface (as sampled by XPS) to that in the bulk using the model of Kirkhof and Moulijn,[24] indicates that the external surfaces of both catalysts are enriched in rhodium. Additionally, analysis of the surface atomic ratios by XPS (Table 2) indicates that this surface enrichment in rhodium occurs in part, during the calcination stage. This may be explained as an effect of capillary flow which drives the solute towards the external surface. Pt deposited as H_2PtCl_6 is expected to be strongly adsorbed at the external surface of the particle[26] (egg shell), and in agreement with XPS results (Table 2) its surface concentration is not dramatically altered following calcination. However, $Rh(NO_3)_3$ is more weakly adsorbed and will be initially adsorbed deeper into the alumina particle.[26] The extent of depletion of this rhodium from the inner structure during heating will depend on the strength of this adsorption and hence on the presence of strong anchorage sites(ie. OH groups). As precalcination at 673 K prior to Rh impregnation [Pt-Rh(step)] is likely to result in the removal of a considerable fraction (ca. 70%)[27] of these, the subsequent adsorption of Rh may be affected. Although rehydration of the alumina surface will occur during aqueous impregnation, it is possible that the resulting surface is modified in some way, producing weaker adsorption sites. In this respect, catalysts prepared by 2-stage impregnation involving acetone,[3] where no rehydration can occur during the second impregnation stage, is likely to strongly effect the resulting catalyst, and may to some extent explain differences between single and two stage prepared Pt-Rh catalysts.[3]

We wish to thank Johnson Matthey for the loan of aqueous rhodium nitrate and for access to the TPR facility. Thanks also to Miss Shaw for her assistance in recording the TPR data and to Prof. R. W. Joyner for helpful discussions regarding the XPS analysis.

References

1 K. C. Taylor in "Catalysis-Science and Technology," ed. J. R. Anderson and M. Boudart, Springer-Verlag, Berlin, 1984, p.121.
2 M. J. D'Aniello Jr., D. R. Monroe, C. J. Carr and M. H Krueger, J. Catal.,1988,109,407.
3 S. H. Oh and J. Carpenter, J. Catal., 1986,98,178.
4 A. G. Bosch-Driebergen, M. N. H. Kieboom, A. Dreumel, R .M. Wolf, F.C.M.J.M. Delft and B.E. Nieuwenhuys, Catal. Letts., 1989,2,235.
5 J. A. Anderson, J. Chem. Soc., Faraday Trans.,1991,87,3907.
6 J. A. Anderson, J. Chem. Soc., Faraday Trans., in press paper 1/05089
7 D. A. Shirley, Phys. Rev.B, 1972,5,4709.
8 J. H. Scofield, J. Electron Spectroscopy and Relat. Phenom. 1976,8,129.
9 M. P. Seah and W. A. Dench, Surf. Interface. Anal., 1979,1,2.
10 R. W. Joyner and B. Williams, manuscript in preparation.

11 J. C. Vis, H. F. J. vant't Blik, T. Huizinga, J. van Grondelle and R. Prins, J. Catal., 1985, 95, 333.
12 H. C. Yao, M. Sieg and H. K. Plummer Jr., J. Catal., 1979, 59, 365.
13 H. Lieske, G. Lietz, H. Spindler and J. Volter, J. Catal., 1983, 81, 8.
14 D. Bazin, H. Dexpert, J. P. Bournonville and J. Lynch, J. Catal., 1990, 123, 86.
15 J. H. A. Martens and R. Prins, Appl. Catal., 1989, 46, 31.
16 J. A. Anderson, M. G. V. Mordente and C.H. Rochester, J. Chem. Soc., Faraday Trans. 1, 1989, 85, 2983.
17 R. W. Broach, J. P. Hickey and G. R. Lester, J. Molec. Catal., 1983, 20, 389.
18 R. F. van Slooten and B. E. Nieuwenhuys, J. Catal., 1990, 122, 429.
19 J. A. Anderson and F. Solymosi, J. Chem. Soc., Faraday Trans., 1991, 87, 3435.
20 B. R. Powell and Y-L. Chen, Appl. Catal., 1989, 53, 233.
21 S. Kim and M. J. D'Aniello, Jr., Appl. Catal., 1989, 56, 45.
22 D. K. Paul and J. T. Yates Jr., J. Phys. Chem., 1991, 95, 1699.
23 M. I. Zaki, T. H. Ballinger and J. T. Yates Jr., J. Phys. Chem., 1991, 95, 4028.
24 F. P. J. M. Kirkof and J. A. Moulijn, J. Phys. Chem., 1979, 83, 1612.
25 L. Lowendahl and J. E. Ottersedt, Appl.Catal., 1990, 59, 89.
26 S-Y. Lee and R. Aris, Catal. Rev.-Sci. Eng., 1985, 27, 207.
27 H. Knozinger and P. Ratnasamy, Catal. Rev.-Sci. Eng., 1978, 17, 31.

TEM and *In Situ* FTIR Characterization of Pd–Fe Particles on Alumina: Geometric and Electronic Effects Due to Iron

E. Merlen, N. Zanier-Szydlowski, P. Sarrazin, and M. L. Hely

INSTUTUT FRANÇAIS DU PÉTROLE, I & 4 AVENUE DE BOIS-PRÉAU, BP 311, 92506 RUEIL-MALMAISON CEDEX, FRANCE

1 INTRODUCTION

The use of bimetallic catalysts at the industrial scale has greatly increased during the last decade. For example, in the fields of catalytic reforming and selective hydrogenation, bimetallic catalysts have enabled significant improvements in catalytic activity and selectivity, as well as increased resistance against coke formation and stability against poisoning by impurities.[1] Recently, significant improvements in catalyst preparation methods have led to a greater control of the supported multimetallic phase.[2,3]

It has been reported in the field of selective hydrogenation that the optimal bimetallic catalyst formulae are obtained for precise particle sizes and compositions depending on the substrate.[4] These results were obtained using metal vapor deposition which enables the production of model bimetallic particles. It is of prime importance to control the particle size and composition to obtain optimal catalytic properties. This cannot be achieved without sophisticated characterization methods, although catalytic testing remains necessary.

The aim of this work is to correlate the catalytic activity of Pd-Fe on alumina for the selective hydrogenation of 1,3-butadiene to butenes, with particle size and composition by systematically investigating the physical properties of these materials. Global and local analytical methods, infrared spectroscopy and electron microscopy, have been used throughout this study. This cross-checked approach appears to be useful for catalytic materials where global characteristics, such as electronic properties, and local characteristics, such as site geometry, are important for the catalytic process. Nevertheless, other analytic methods, particularly X-ray fluorescence and XPS, must be used to completely describe these materials. Finally, the catalysts have been tested for 1,3-butadiene hydrogenation activity and selectivity to correlate their catalytic properties with the characterization results.

2 EXPERIMENTAL

Catalysts Preparation

The catalysts were prepared using γ_t-Al_2O_3 as a support (surface area = 150 m^2/g and total pore volume = 1.05 cm^3/g) which was in powder form and calcined at 573 K prior to impregnation. The impregnation of palladium was achieved by contacting the support with a solution of bis-acetylacetonatopalladium (Pd(acac)$_2$) in toluene (2.6 g of Pd/l). The volume of the solution was 5 times the pore volume of the alumina. After 72 hours of contact, the powder was filtered off, dried and calcined in air for 2 hours at 573 K. The monometallic catalysts were reduced under a hydrogen flow for 2 hours at 723 and 573K respectively depending on the desired dispersion (Table 1).

The bimetallic catalysts were prepared by contacting the reduced monometallic catalysts with a solution of iron pentacarbonyl (Fe(CO)$_5$) in toluene. The amount of iron introduced was that necessary to obtain an atomic iron to accessible palladium ratio of either 0.5 or 2.0. Subsequently, the mixture was heated from ambient temperature to 373 K (0.5 K/min) under an argon flow. Iron pentacarbonyl decomposition was monitored by gas chromatographic analysis of the CO content in the argon flow leaving the reactor. After decomposition of Fe(CO)$_5$, the solvent was removed by evaporation and the catalyst was washed with fresh toluene, dried and then calcined for 2 hours at 573 K. The bimetallic catalysts thus obtained were reduced for 2 hours at either 573 or 803 K. The palladium contents of the samples were 1.3 %wt for series A and 1.25 %wt for the series B while the iron content ranged from 0.15 %wt to 0.86 %wt (Table 1).

The metal surface areas were measured by CO chemisorption under dynamic conditions at room temperature.[5] The catalyst samples were reduced under hydrogen at 573 K for 2 hours and desorbed under argon at 573 K for 2 hours as well. The samples were then cooled under argon and known volumes (0.5cm^3) of CO were injected. A stoichiometry of one CO molecule per atom of palladium was used to determine the number of exposed Pd atoms. This stoichiometry has been reported to be valid for alumina supported monometallic catalysts over a large range of particle size.[6]

Electron Microscopy

Samples were crushed and deposited on holey carbon microscopy copper grids. All specimens were examined by conventional transmission electron microscopy (CTEM) using a JEOL 120CX at 100 kV. Analysis was performed using a VG HB5 dedicated STEM (scanning transmission electron microscope) equipped with a Tracor TN5500 energy-dispersive X-ray spectrometer (EDX).

Table 1 Physico-chemical characteristics of the samples

Samples	Reduction temperature K ***	Pd %wt*	Fe %wt*	Pd surface exposed atoms ** %(or dispersion)
1A		1.30	0	38
2A	573	1.30	0.15	37
3A	803	1.30	0.15	38
4A	573	1.30	0.47	25
5A	803	1.30	0.47	22
1B		1.25	0	97
2B	573	1.25	0.39	51
3B	803	1.25	0.39	35
4B	573	1.25	0.86	38
5B	803	1.25	0.86	22

* as determined by X-ray fluorescence spectroscopy
** as determined by CO chemisorption
*** reduction temperature of the bimetallic catalysts

Infrared Cell Description and Operating Conditions

The chemisorption of CO was carried out in an *in-situ* dynamic infrared cell. This high temperature infrared cell, described elsewhere,[7] has been installed in a DIGILAB FTS80 sample compartment. The samples were pressed into self-supporting disks, 16 mm diameter, 0.2 mm thick, weighing 10-15 mg. Before spectra acquisition, the catalysts were activated overnight under a helium flow at 373 K to eliminate water. After reduction of the catalysts under hydrogen for 4 hours at 573 K or 803 K, helium was admitted for 2 hours at the same temperature. The temperature was then lowered to ambient and 500 microliters of CO were introduced. The gas flow rate was 0.5 l/h. One spectrum per minute was recorded by averaging 30 interferograms with a resolution of 4 cm^{-1}. CO adsorption and desorption were followed continuously for 30 minutes. After subtraction of CO in the gas phase, the surface areas of the chemisorbed species between approximately 2120 and 1700 cm^{-1} were evaluated. The spectrum corresponding to the maximum chemisorption was retained for this work.

Catalytic Tests Procedure

The catalytic activities of the samples have been evaluated for the selective hydrogenation of 1,3-butadiene to 1-butene and 2-butenes in heptane. The tests were carried out in the liquid phase in a continuous stirred tank reactor (CSTR) at a constant pressure of 1 MPa. The hydrogenation rate was followed by recording the pressure drop in a constant volume hydrogen ballast upstream of the pressure regulator as a function of time. Cooling water was circulated through the reactor jacket in order to maintain a constant temperature of 293 K. The

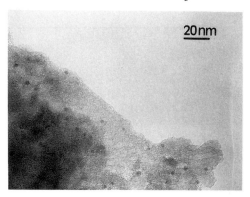

Figure 1   Electron micrograph obtained for sample 3A

hydrocarbon samples were analyzed with an Intersmat IGC 112F chromatograph equipped with a flame ionization detector using a 6 mm diameter ββ' ODPN phase supported on chromosat PAW column at room temperature.

3   RESULTS

Electron Microscopy

Metallic Phase Dispersion. Typical micrograph obtained on sample 3A is shown in Figure 1. Histograms were determined from a large number of particles and the average sizes are reported in Table 2. When the dispersion of the monometallic precursor Pd/Al_2O_3 is low (series A), the addition of Fe does not modify the dispersion regardless of the preparation conditions considered in this work. For series B, the monometallic precursor is highly dispersed: no metal particles are observed, indicating that the particle sizes are less than 0.7 nm which is the detection limit with alumina support. The introduction of the second metal leads to a sintering of the metallic phase independent of the experimental conditions (see series B in Table 2).

Individual Particle Composition. Four samples (3A, 4A, 5A and 5B) were selected to be analyzed by STEM. The amount of iron measured in the particles accounts for about 40 % of the iron introduced during synthesis. A large part of Fe was probably "hyperdispersed" on the support. The amount of iron in the particles did not depend on the reduction temperature.

Infrared Spectroscopy

As the amount of iron is very small, the infrared

spectrum of CO chemisorbed on it is not observed[8] and the measured infrared spectra are characteristic of CO adsorbed on palladium which is electronically and/or geometrically modified by iron. It is possible to analyze the evolution of wavenumber shifts and relative intensity variations of the bands corresponding to CO adsorbed on one or several palladium atoms ($\bar{\nu} > 2000 cm^{-1}$, linear species; $\bar{\nu} < 2000 cm^{-1}$, bridged species)[9] with and without the addition of iron. In order to avoid the effects of dipole-dipole interactions,[10] the retained IR spectra were recorded at maximum CO coverage. Wavenumber shifts of linear forms were correlated with electronic effects.[11] A decrease of the absorbance peak areas of bridged species with respect to linear species is attributed either to dilution effects for bimetallic catalysts[12] or to an increase of the dispersion for monometallic catalysts.[13]

Number and Positions of Absorption Bands. The infrared spectra obtained at maximum chemisorption of CO on the mono- and bi-metallic catalysts are shown in Figure 2. The absorption bands near 2200 cm^{-1} correspond to CO chemisorbed on alumina. In Table 2, the wavenumbers of the absorption bands of CO on one or several palladium atoms and the variation of the wavenumber of linear species of the bimetallic catalysts with respect to the wavenumber of the linear species of monometallic catalyst having the same average particle size (noted $\Delta\bar{\nu}_L$) are presented. For series B, the $\bar{\nu}_L$ for a monometallic Pd catalyst with an average particle size of 1.8 nm has been interpolated to be 2083 cm^{-1} from a study of $\bar{\nu}_L$ as a function of the monometallic particle size. For the bimetallic catalysts, a doublet near 2300-2400 cm^{-1} is observed corresponding to CO_2 resulting from either CO dismutation or additional reduction of iron from Fe^{2+} state to Fe^0. Indeed, Fe^{2+} detected by XPS (E_L=712 eV) on samples 4B and 5B respectively reduced at 573K and 803K should exist on the support near the bimetallic particles under oxide or aluminate states. The shift in the linear CO stretching vibration, $\Delta\bar{\nu}$, varies from -4 to +5 cm^{-1} for series A and from -4 to +12 cm^{-1} for series B. The linear stretching band always exhibits a shoulder at lower wavenumbers. The band characteristic of the bridged species is split into two bands except for sample 5A and the series B samples.

Variation of the Ratio of the Area of the Linear Species to the Bridged Species (L/B). After reduction of bimetallic catalysts (2A, 4A) at 573 K, the ratio L/B increases slightly (Table 2). However, reduction at 803 K, samples 3A and 5A, leads to an L/B increases from 0.4 for the monometallic catalyst 1A to 1.7-1.9 for the bimetallic catalysts. A reduction temperature of 573 K does not alter the L/B ratio of 1.1 for series B catalysts. The ratio increases when the higher reduction temperature is employed (803 K) and is higher for high iron contents.

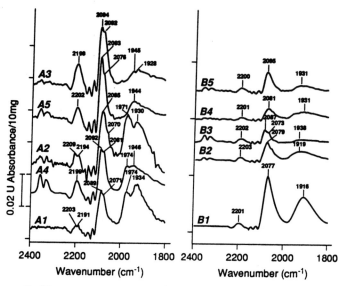

Figure 2 IR spectra at maximum CO chemisorption on series A and B

Table 2 Catalytic activity and Infrared and TEM characteristics of the samples

Samples	$\bar{\nu}_L$ (cm^{-1})	$\Delta\bar{\nu}_L$ (cm^{-1})	$\bar{\nu}_B$ (cm^{-1})	L/B	TON (s^{-1})	particle size** (nm)
1A	2089 2071		1974 1934	0.4	12.7	3.0
2A	2085 2070	-4	1971 1930	0.5	13.4	3.0
3A	2094 2082	5	1945 1928	1.9	14.6	3.0
4A	2092 2081	3	1974 1946	0.8	13.1	3.0
5A	2093 2076	4	1944	1.7	17.3	3.0
1B	2077		1916	1.1	5.5	<0.7
2B	2079 2063	-4*	1919	0.8	9.3	1.8
3B	2087	4*	1938	1.6	16.1	1.8
4B	2081	-2*	1931	1.1	11.4	1.8
5B	2095 2073	12*	1931	2.3	15.4	2.1

* calculated from $\bar{\nu}_L$ of Pd catalyst with the same particle average size ($\bar{\nu}_L = 2083$ cm^{-1})
** average size as determined by TEM

Catalytic Tests

The turnover number (TON) measured with each catalyst is reported in Table 2. It appears, for series A, that iron addition changes slightly the TOF in comparison with the monometallic catalyst 1A. This effect is more important for the two bimetallic catalysts 3A and 5A which have been reduced at 803 K. The effect of iron is much more pronounced for series B. Catalysts 2B and 4B (reduced at 573 K) exhibit TONs which are twice the TON of catalyst 1B. Again, a positive effect of the reduction temperature is observed. Catalysts 3B and 5B (reduced at 803 K) yield TONs which are three times that of catalyst 1B. The initial selectivities regarding 1,2-addition and 1,4-addition of hydrogen and the trans/cis distribution of the 2-butenes have also been determined. No variation of these values has been observed as a result of modification of the palladium particles by iron.

4. DISCUSSION

Bimetallic Particle Size

The average particle size determined by TEM clearly demonstrates that the bimetallic particle size depends only on the initial palladium particle size. The amount of iron and the reduction temperature do not influence the final size of the bimetallic particle. The particle size variation between the monometallic and the bimetallic catalysts increases with the dispersion of the monometallic precursor. For series A (initial dispersion = 38 %), there is no size variation between the monometallic and the bimetallic catalysts; but, for series B (initial dispersion = 97 %), the particle size varies from lower than 0.7 nm for sample 1B to 1.8-2.1 nm for the corresponding bimetallic catalysts (Table 2). It has been verified with sample 2B that the particle size was the same before and after reduction. There is no redispersion of the metallic phase by reaction of the Pd aggregates with carbonyl iron under these experimental conditions. Thus, it can be proposed that the bimetallic catalyst is sintering to a given size upon the addition of iron, independent of the iron content and the reduction temperature.

Iron Distribution on Pd-Fe Particle Surface

The average bimetallic particle size does not vary with the iron content or the reduction temperature for series A (Table 2). On the other hand, the number of accessible palladium atoms (Table 1) decreases with increasing iron content. STEM analysis show that the particles are bimetallic and contain about 40 % of the introduced iron. The surface iron contents of bimetallic catalysts 4A and 5A are higher than those of catalysts 2A and 3A. Moreover for the same quantity of iron the ratio L/B increases with increasing reduction temperature while dispersion remains constant. We can thus conclude that for the same quantity of iron the distribution of iron atoms on the surface of the bimetallic particles is more homogeneous after reduction at 803 K. For the same average bimetallic particle size in series B, the dispersion decreases with increasing iron content and reduction temperature. The L/B ratio increases with the reduction temperature. Similarly we conclude that there is a more homogeneous distribution of iron on the palladium particle surface and that the surface is iron enriched. A schematic presentation of the bimetallic catalyst surfaces is given in Figure 3.

The iron distribution on the surface depends on the reduction temperature of the bimetallic catalyst, i.e. the more homogeneous distribution is obtained for the higher reduction temperature. The quantity of iron introduced is related to the monometallic particle size: the smaller the monometallic particle size, the higher the quantity of introduced iron. The bimetallic particle surface of sample 5B is richer in iron than that of sample 5A reduced at the same temperature. Both samples have the same dispersion (22%). Sample 5B exhibits a lower dispersion than expected considering its lower average particle size. This is explained by a higher iron content on the surface.

Electronic Effect of Iron on Palladium

An electronic effect was observed for sample 5B which showed shift of the linear CO wavenumber. $\Delta\bar{v}_L$ is equal to +12 cm^{-1}. This variation corresponds to an increase of 1.2% of the force constant between C and O atoms therefore to a decrease in the $d\pi \rightarrow \pi^*$ metal to ligand backdonation[14] due to the addition of iron. This is in agreement with the electronic configurations of Pd^0 ($4d^{10}$) and Fe^0 ($3d^8$) which suggest that Fe^0 should be more electron acceptor enabling it to attain the half-filled orbital configuration.

Correlation with Catalytic Activity

The 1,3-butadiene hydrogenation selectivity is only slightly affected by iron addition to the monometallic catalyst. However the TON decreases with increasing monometallic catalyst dispersion. In other words, the activity increases with the reduction temperature and with a better distribution of iron on the bimetallic particle surface. The evolution of the catalytic activity correlates with a particular distribution of iron on the bimetallic particle surface (Figure 3). The geometric effect (dilution and size effects) appears to be a very important parameter. The same is true for the quantity of iron. The effect is particularly noticeable in series B. The TONs of series B bimetallic catalysts are comparable

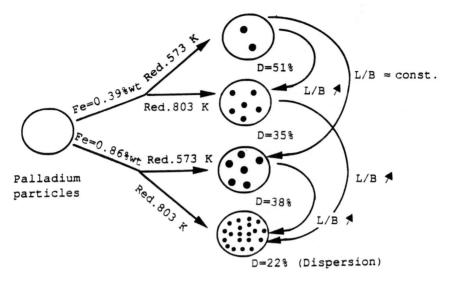

Figure 3 Schematic representation of the bimetallic catalyst surfaces for series B (the same presentation is valid for series A): particle size does not depend on iron content or reduction temperature (CTEM results) and iron distribution on the palladium particle surface depends on the reduction temperature (IR(CO) and chemisorption results). Iron is represented as dark disks

to those of series A with a lower dispersion. It has been found that the TON increases with the particle size on monometallic Pd catalysts.[6] But the sintering involved in the series B is not enough to explain the TON variation observed. Only an electronic effect of iron on palladium can account for the comparable activity with a lower average particle size.

5 CONCLUSION

This work has enabled us to describe Pd-Fe dispersions on alumina catalysts prepared by Fe(CO)$_5$ decomposition. The results show that the presence of iron alters the geometric and electronic nature of the palladium particles. Sintering of Pd-Fe bimetallic particles, resulting from a "hyperdispersed" Pd precursor, partially explains the increased hydrogenation activity of these catalysts with respect to the corresponding Pd catalyst. Homogeneous distribution of iron on the metallic particle surface appears to correlate with catalytic activity. It is difficult to differentiate the importance of the electronic and geometric effects on the catalytic activity.

REFERENCES

1. J.P. Boitiaux, J. Cosyns, M. Derrien and G. Léger, Hydrocarbon Processing, 1985, March, 51.
2. M. Che and C.O. Bennett, Adv. Catal., 1989, 36, 55.
3. R.L. Moss, Catalysis, 1981, 4, 31.
4. V. de Gouveïa, PhD Thesis, Université Pierre et Marie Curie, (Paris VI) 1988.
5. C.S. Brooks and V.J. Kehrer, Anal. Chem., 1969, 41, 103.
6. J.P. Boitiaux, J. Cosyns and S. Vasudevan, in Preparation of Catalysts III, G. Poncelet, P. Grange and P.A. Jacobs (Editors), 1983 Elsevier, Amsterdam.
7. J.F. Joly, N. Zanier-Szydlowski, S. Collin, F. Raatz, J. Saussey and J.C. Lavalley, Catalysis Today, 1991, 9, 31-38.
8. A.F.H. Wielers, A.J.H.M. Kock, C.E.C.A. Hop and J.W. Geus, J.Catal., 1988, 117, 1.
9. M.A. Chesters, G.S. Mc Dougall, M.E. Pemble and N. Sheppard, Surf.Sci., 1985, 164, 425-436.
10. C. Binet, A. Jadi and J.C. Lavalley, J.Chem.Phys., 1989, 86(3).
11. M.C. Kung and H.H. Kung, Catal.Rev.Sci.Eng., 1985, 27(3), 425-460.
12. M. Primet, M.V. Mathieu, W.M.H. Sachtler, J.Catal., 1976, 44, 324-327.
13. J.L. Duplan, M. Praliaud, Appl.Catal., 1991, 67, 325-335.
14. L.H. Little,'Infrared Spectra of Adsorbed Species' 1966, Academic Press, London.

ACKNOWLEDGMENTS

We thank the staff of IFP's Physical Chemistry Department for its contribution. We especially express our thanks to Ms. V. Poitrineau for STEM analysis, Mr. B. Moisson for IR experiments and Mr. C.T. Nguyen for his help in catalyst preparing and testing.

In Situ Solid-state NMR Study of the Adsorption of Methanol and Water on the Cu/ZnO/Al$_2$O$_3$ Methanol Synthesis Catalyst

A. Bendada, J. B. C. Cobb, B. T. Heaton, and J. A. Iggo
DEPARTMENT OF CHEMISTRY, UNIVERSITY OF LIVERPOOL, LIVERPOOL L69 3BX, UK

1 ABSTRACT

The adsorption of methanol and water on Cu/ZnO/Al$_2$O$_3$ methanol synthesis catalysts has been studied over the temperature range of 293-480 K, using *in-situ* ^{1}H and ^{2}H NMR spectroscopies. Methanol adsorbs to give initially hydrogen gas, shown by deuterium labelling to come from the OH(D), and methoxy groups; then weakly physisorbed methanol which can be readily removed on evacuation at 293 K. At high coverage a second, more tightly bound physisorbed species is observed. Temperature programmed reaction (TPR) of methanol adsorbed on the catalyst reveals the onset of dehydrogenation of the methoxy groups at 333 K. This reaction has gone to completion at 480 K when only gaseous hydrogen is seen. After TPR the catalyst no longer adsorbs hydrogen, methanol or water. Water adsorbs in a similar manner to methanol. Hydrogen gas is seen early in the reaction followed by the appearance of weakly adsorbed water.

2 INTRODUCTION

Details of adsorption processes, the structure of catalysts and the identity of the surface adsorbates are essential to the understanding of the mechanisms of catalytic processes. Advances in acquiring this information have been achieved by solid-state nuclear magnetic resonance in several systems. One common feature of all these studies is that samples are sealed after preparation/activation and before commencing the NMR experiment. It is thus not possible to vary the conditions other than temperature during the NMR experiment nor is it possible to admit other reagents to the reactor. Inevitably there must be a considerable delay between admitting the adsorbing reactant and commencing the NMR experiment. Fast, initial reactions cannot therefore be observed.

We have previously described a multinuclear NMR cell/microreactor for the *in-situ* study of surface adsorbates on supported metal catalysts.[1,2] One feature of this cell is that the reactor remains connected to the gas handling lines. It is therefore possible to evacuate or admit reactive gases/vapours to the cell during the NMR experiment. Adsorption reactions can therefore be studied in real time rather than reconstructing the reaction from samples hopefully quenched and sealed at different stages of reaction. We now report the results of our investigation of the adsorption and subsequent reaction of methanol and water on $Cu/ZnO/Al_2O_3$ methanol synthesis catalysts.

An *ex-situ* ^1H and ^{13}C NMR spectroscopic study on the adsorption of hydrogen and methanol on copper/zinc oxide/ alumina methanol synthesis catalysts has appeared.[3]

3 EXPERIMENTAL

NMR spectra were obtained on a Bruker WM200 WB spectrometer, operating field 4.7 T, using the probe described previously,[1] operating at 200 and 30.778 MHz for ^1H and ^2H respectively. The FID's were accumulated using 18 μs, 45°, pulses and a delay of 0.3-0.5 s between successive pulses over a sweep width of 60 kHz (^1H) and 30 kHz (^2H). Typically 100-1000 (^1H) and 4000-10000 (^2H) scans were acquired. Spectra were referred to $H_2(g)$ = 0 ppm (^1H) and $D_2(g)$ = 0 ppm (^2H). Copper based methanol synthesis catalysts were supplied by ICI Katalco; ZnO/Al_2O_3 (80/20 wt%) was prepared by co-precipitation of basic salts of zinc and aluminium from approximately 1 M nitrate solution as described in the literature.[4] Powdered catalyst samples (1-2 g) were activated in flowing H_2, H_2/N_2 or CO (5 lh^{-1}) at 493 K overnight in the NMR probe. Pressurisation, evacuation to a pressure of 10^{-4} Torr and dosing with vapour of probe molecules were performed *via* the gas handling lines described previously.[2]

Industrial grade hydrogen (99.5 %) (BOC) was used for activation purposes only. CP grade hydrogen (99.999 %) (BOC) was used without further purification. CH_3OH (BDH), CH_3OD, CD_3OH and CD_3OD (99.5 atom% D) (Goss) were dried by refluxing over activated magnesium overnight under nitrogen atmosphere and immediately loaded into stainless steel vessels under nitrogen atmosphere, connected to the gas handling lines and the connections purged of air.

4 RESULTS

Adsorption of CH₃OD.

The ^{2}H NMR spectrum of Cu/ZnO/Al$_2$O$_3$ exposed briefly to CH$_3$OD shows the presence of a sharp peak at 0 ppm, half height linewidth (W$_{1/2}$) = 4 ppm and a very weak broad peak at 22 ppm, W$_{1/2}$ = 100 ppm. On further exposure of the catalyst to CH$_3$OD, the intensity of the sharp peak increases slightly then remains roughly constant whereas the broad peak increases significantly in intensity. These resonances are completely lost on evacuation at 293 K, figure 1.

Adsorption of D₂O.

The admission of D$_2$O into the reactor containing catalyst which had been exposed briefly to CH$_3$OD and evacuated as described above resulted in the appearance of a sharp resonance at 0 ppm, W$_{1/2}$ = 4 ppm, and a broad peak at 12 ppm, W$_{1/2}$ = 125 ppm, in the ^{2}H NMR spectrum, figure 2. Evacuation of the system resulted in the removal

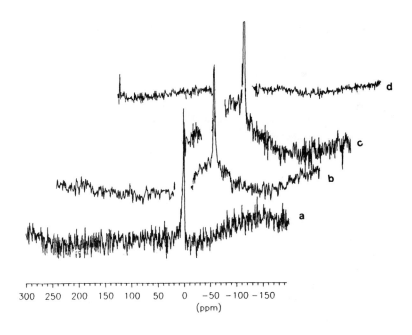

Figure 1. *In-situ* ^{2}H NMR spectra of the reaction between CH$_3$OD and Cu/ZnO/Al$_2$O$_3$ at 293 K with time; a) t = 1 min; b) t = 5 min; c) t = 10 min; d) after evacuation.

Table 1. ^{2}H chemical shifts/ppm and linewidths/ppm of resonances recorded upon exposure of Cu/ZnO/Al$_2$O$_3$ to methanol and water.

Adsorbate	Shift	Linewidth	Assignment	Shift	Linewidth	Assignment
CH$_3$OD[a]	0	4	D$_2$(g)	22	100	CH$_3$OD(ads)
D$_2$O[a]	0	4	D$_2$(g)	12	125	D$_2$O(ads)
D$_2$O[e]	0	250	OD(ads)	-	-	-
CD$_3$OH[a]	8	35	CD$_3$O(ads)	-	-	-
CD$_3$OH[b]	8	35	CD$_3$O(ads)	0	4	D$_2$(g)
CD$_3$OH[c]	0	4	D$_2$(g)	12	120	D$_2$CO(ads)
CD$_3$OH[d]	0	4	D$_2$(g)	-	-	-
CD$_3$OD[a]	0	32	CD$_3$O(ads)	12	102	D$_2$CO(ads)
CD$_3$OD[b]	0	4	D$_2$(g)	12	100	D$_2$CO(ads)
CD$_3$OD[c]	0	4	D$_2$(g)	-	-	-
CD$_3$OD[e]	-	-	-	0	240	OD(ads)

(a) T = 293 K, (b) T = 333 K, (c) T = 410 K, (d) T = 453 K, (e) evacuated at 293 K

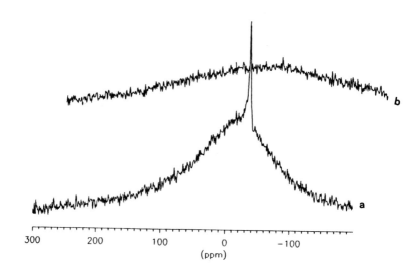

Figure 2. *In-situ* ^{2}H NMR spectra of the reaction between D$_2$O and Cu/ZnO/Al$_2$O$_3$ at 293 K; a) admission of D$_2$O; b) evacuated.

of these resonances and allowed the observation of a very broad peak around 0 ppm, $W_{1/2}$ = 250 ppm.

Adsorption of CD<u>3OH.</u>

The adsorption of CD_3OH on a fresh sample of activated $Cu/ZnO/Al_2O_3$ gives initially a broad peak at 8 ppm, $W_{1/2}$ = 35 ppm in the ^2H NMR spectrum and a weak resonance at 0 ppm in the ^1H NMR spectrum, figures 3,4. On increasing exposure to CD_3OH the intensity of the deuterium spectrum increases whereas that of the proton NMR spectrum remains unchanged, figure 5.

Temperature programmed reaction of adsorbed C<u>D3OH.</u> The subsequent reaction with increasing temperature of the sample was monitored by ^2H NMR spectroscopy, figure 5. Reaction commenced at *ca.* 333 K as evidenced by the appearance of a sharp resonance at 0 ppm, $W_{1/2}$ = 4 ppm. This resonance increases in intensity as the intensity of the original broad resonance decreases. A new, very broad resonance is evident at around 413 K. At 453 K, only the sharp line is seen.

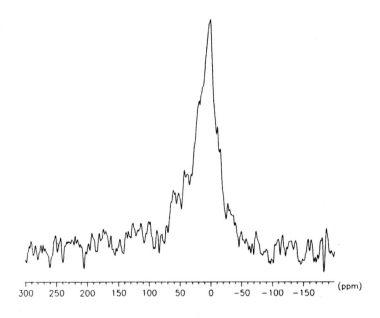

Figure 3. *In-situ* ^2H NMR spectrum of the reaction between CD_3OH and $Cu/ZnO/Al_2O_3$ at 293 K.

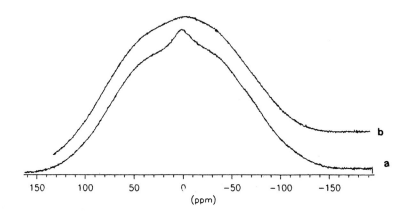

Figure 4. *In-situ* ^{1}H NMR spectra of the reaction between CD_3OH and $Cu/ZnO/Al_2O_3$; a) before TPR b) after TPR, cooling to 293 K and evacuation.

Cooling the system to 293 K had no effect on the observed ^{2}H NMR spectrum. On brief evacuation at 293 K the sharp resonance disappeared. The ^{1}H NMR spectra recorded prior to or after heating showed only the presence of the very weak resonance at 0 ppm described in the previous section. This resonance is completely removed on evacuation at 293 K, figure 4.

Adsorption of CD_3OD.

On admission of a small quantity of CD_3OD to activated $Cu/ZnO/Al_2O_3$ the ^{2}H NMR spectrum shows an asymmetric peak at 0 ppm, $W_{1/2}$ = 35 ppm growing out of a broad resonance. With further exposure of the catalyst to CD_3OD, the ^{2}H NMR spectrum reveals the presence of two overlapping peaks, figure 6. with linewidths *ca.* 4 and 35 ppm. The species present at 293 K appear to react on heating the catalyst in a similar manner to that observed with CD_3OH. The strong peak became sharper but the broad peak became very weak and disappeared at 480 K. Cooling of the reactor to 293 K had no observable effect upon the final spectrum recorded at 480 K. The ^{2}H NMR spectrum of the catalyst briefly evacuated at 293 K shows the presence of the sharp peak at 0 ppm and a broad peak around 0 ppm with linewidths of 4 ppm and 240 ppm respectively. On further evacuation the sharp resonance disappears.

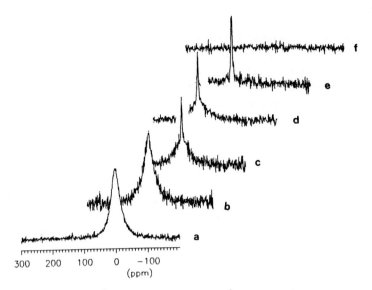

Figure 5. *In-situ* ^{2}H NMR spectra of temperature programmed reaction of CD$_3$OH on Cu/ZnO/Al$_2$O$_3$; a) 293 K; b) 333 K; c) 363 K; d) 413 K; e) 453 K; f) after evacuation at 293 K.

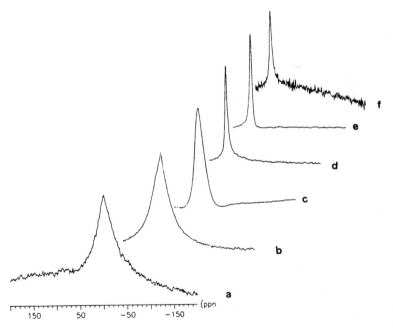

Figure 6 *In-situ* ^{2}H NMR TPR spectra of CD$_3$OD on Cu/ZnO/Al$_2$O$_3$; a) 293 K; b) 293 K, higher coverage; c) 293 K, saturation; d) 460 K; e) 293 K; f) 293 K, brief evacuation.

5 DISCUSSION

The adsorption of methanol on $Cu/ZnO/Al_2O_3$ as revealed by *in-situ* ^1H and ^2H NMR spectroscopy occurs in three stages. Initially a small amount of methanol is seen to dissociatively adsorb giving gaseous hydrogen (derived from the hydroxyl group of the methanol as evidenced by the sharp resonance at 0 ppm in ^2H NMR spectrum figure 1a) and chemisorbed methoxy groups. Subsequently, weakly physisorbed methanol, giving a relatively immobile OH group, is seen as evidenced by the broad line in figure 1b. This physisorbed methanol is readily lost on evacuation, figure 1d. This interpretation is consistent with the observations of Borisov et al.[5] and Miyazaki and Yasumori,[6] who report that methanol weakly adsorbs on copper and of Bowker and Madix,[7] and Sexton[8] who demonstrated that the hydroxyl proton of methanol is readily lost below 373 K. Hence the initial adsorption reactions may be shown pictorially as in figure 7.

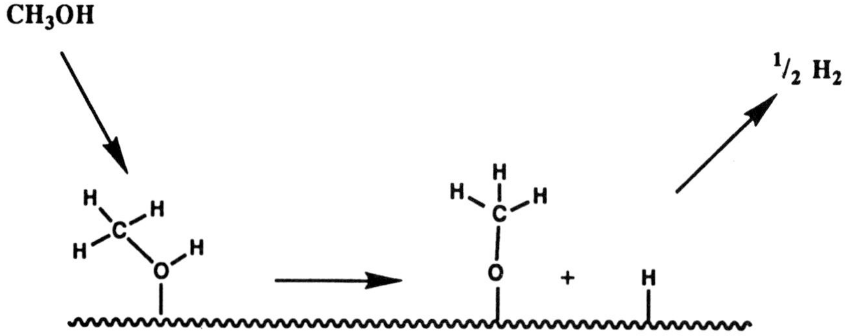

Figure 7. Adsorption of methanol on $Cu/ZnO/Al_2O_3$.

One point of interest in our results is that the hydroxyl hydrogen is lost into the gas phase. A small amount of Knight shifted hydrogen on copper was observed on initial exposure of the catalyst to methanol. The intensity of this peak did not increase with increasing exposure.[9] We attribute this to dissociative chemisorption of methanol in which both the hydrogen and methoxy groups initially occupy the same copper sites. As coverage increases the methoxies remain on the surface whilst the hydrogen is liberated into the gas phase. This contrasts with the report of Dennison et al.[3] who found hydrogen on copper on all samples exposed to methanol. At high coverage a second physisorption site for methanol is occupied. Methanol adsorbed on this site is lost only slowly on evacuation.

The TPR spectra are consistent with dehydrogenation of the methoxy groups commencing around 333 K in agreement with Bowker et al.'s TPD results which showed co-incident desorption of formyaldehyde, methanol and hydrogen at 340 K.[10] Thus the evolution of D_2 is seen in the ^2H NMR spectrum, figure 5. Concomitant with this there is a decrease in the intensity of the resonance of the CD_3O groups and the appearance of a very broad resonance assigned to chemisorbed formyl or formate. Eventually only D_2 is seen in the ^2H NMR spectrum implying either complete dehydrogenation of methanol to carbonate or formation of a very tightly bound, immobile, formate. The ^1H NMR spectra taken before and after TPR are broadly similar showing only the H_2 derived from the OH group of methanol. On evacuation of the reactor all NMR resonances other than that of OD(ads) disappear, consistent with the spectra after TPR being due to $D_2(g)$ and $H_2(g)$. The dehydrogenation products remain on the surface of the catalyst since the catalyst is now unable to adsorb H_2 or methanol.

Water adsorbs on the active catalyst in a similar manner to methanol. Thus the evolution of $D_2(g)$ is seen immediately upon admission of D_2O together with a broad resonance of weakly physisorbed D_2O, figure 2. On evacuation these resonances disappear allowing the observation of chemisorbed, immobile OD, the resonance of which was previously obscured by the narrower, more intense resonances of $D_2(g)$ and weakly physisorbed D_2O.

6 CONCLUSIONS

This study has traced the back reactions of methanol to carbonate, and water to surface hydroxyl on $Cu/ZnO/Al_2O_3$. In both cases the hydrogen formed appears to desorb readily from the surface at both 293 K and at 453 K indicating substantial oxidation of the copper surface or competition for adsorption sites on the copper since we and others[1,3] have shown that hydrogen readily adsorbs at 293 K on the metallic copper of reduced $Cu/ZnO/Al_2O_3$.

The surface carbonate formed in the methanol decomposition reaction poisons both the hydrogen adsorption and the methanol chemisorption sites. This may indicate a close connection between the hydrogen and methoxy adsorption sites. Similarly we have not seen hydrogen adsorbed on copper on working catalysts (catalysts run on stream (syngas 50 bar, 500 K)) or on such catalysts after cooling to 293 K.

On these last a resonance assigned to methoxy or formate is observed after evacuation of the reactor.[9]

These observations are consistent with the ICI view of methanol synthesis on copper catalysts.[11] Reaction occurs on the copper surface although our experiments suggest that the availability of hydrogen on such a surface will be low due to partial oxygen coverage of the surface and competition with more strongly bound species for adsorption sites.

7 ACKNOWLEDGEMENTS

We thank the SERC for financial support, Dr. N.J. Clayden for helpful discussions and Mr. G. Chinchen for catalyst samples and helpful discussions.

REFERENCES

1. A. Bendada, G. Chinchen, N.J. Clayden, B.T. Heaton, J A. Iggo and C.S. Smith, Catal.Today, 1991, **9**, 129.
2. A. Bendada, J.B.C. Cobb, B.T. Heaton and J.A. Iggo, accepted for publication.
3. P.R. Dennison, K.J. Packer and M.S. Spencer, J.Chem.Soc., Faraday Trans., 1989, **85**, 3537.
4. R.G. Herman, K.S.G.W. Klier, B.P. Finn and J.B. Bulko, J.Catal, 1979, **56**, 407.
5. A. S. Borisov, V. V. Tsvetkov and V. D. Yagodovskii, Russ.J.Phys.Chem., 1976, **50**, 1288.
6. E. Miyazaki and I. Yasumori, Bull.Chem.Soc.Jpn., 1967, **40**, 2012.
7. M. Bowker and R. J. Madix, Surf.Sci., 1980, **95**, 190
8. B. A. Sexton, Surf.Sci., 1979, **88**, 229.
9. A. Bendada, J.B.C. Cobb, B.T. Heaton and J.A. Iggo, unpublished results.
10. M. Bowker, R.A. Hadden, H. Houghton, J.N.K. Hyland and K.C. Waugh, J. Catal., 1988, **109**, 263.
11. G.C. Chinchen, K. Mansfield and M.S. Spencer, Chemtech, 1990, 92.

The Catalytic Decomposition of Formic Acid by the Cu/Pd [85:15]{110} P(2X1) Surface

Mark A. Newton, Stephen, M. Francis, and Michael Bowker
SURFACE SCIENCE RESEARCH CENTRE AND LEVERHULME CENTRE,
UNIVERSITY OF LIVERPOOL, LIVERPOOL L69 3BX, UK

ABSTRACT

The catalytic decomposition of formic acid by the CuPd[85:15]{110} surface has been studied over a range of pressures. The surface has been found to dehydrogenate formic acid giving as products CO_2 and H_2. The activation energy for the overall reaction is ~36($\pm$6) kJmol^{-1}, while TPD shows a formate species to be the most stable surface intermediate with a decomposition energy of ~ 125($\pm$3) kJmol^{-1}. The results are compared to those deduced for single crystal Cu{110}. Further, a poisoning of the decomposition reaction is found for cases when the sample has been left in the high pressure cell for some time before the reaction is attempted. This effect, though decreasing the observed rate of decomposition considerably, has little effect on the apparent activation energy.

1 INTRODUCTION

The "pressure gap" that exists between studies on real catalysts and those made on well defined single crystal surfaces under U.H.V conditions has always been a contentious issue. In recent years significant advances have been made toward closing this gap, through the use of high pressure facilities directly attached to U.H.V systems. These allow reactions to be carried out over a wide range of presures well as providing pre and post reaction U.H.V surface analysis in many cases.

Formate intermediates have been proposed as a direct participant in several important catalytic reactions and particularly in the Cu catalysed synthesis of methanol [1]. A knowledge of the parameters defining the binding of this intermediate to the surface is important for the understanding of the whole reaction scheme in which it is involved. The catalytic decomposition of formic acid by metal surfaces provides a relatively simple model reaction to perform, and in conjunction with techniques such as thermal desorption spectroscopy (TDS), provides a means to elucidate these kinetic parameters.

The effects of alloying on catalytic reactivity are important and potential sources of new or improved activity. The alloy under study here forms a p(2x1) ordered surface phase under thermal treatment (<670K)[2,3]. The composition and structure of this surface has been determined via XPS, LEIS, LEED [3]. The deduced structure is shown in fig 1. The top layer comprises an all Cu layer, the second an ordered layer of CuPd. The Pd in the second layer will not be able to bond directly to adsorbed species, and it is not easy to see how an ensemble effect is possible at this surface. Differences between the reactivity of this surface and the analogous Cu surface, therefore, must arise from Pd induced electronic peturbation of the surface and as such provide a unique example of a ligand effect at an alloy surface.

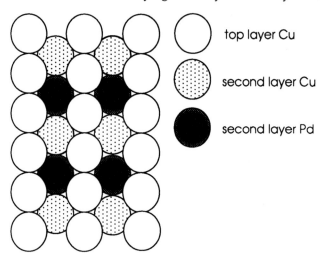

Fig 1. The structure of the CuPd [85:15] {110} p(2x1) surface

This paper seeks to detail the interaction of HCOOH with this surface under UHV and high pressure conditions. A quantification of the nature and size of the ligand effect due to alloying in the decomposition of the formate intermediate is also presented.

2 EXPERIMENTAL

The experiments were performed in a three stage vacuum chamber comprising a preparatory chamber, analysis chamber and high pressure cell. The preparatory chamber was used for the cleaning of the sample via Ar^+

bombardment. Preparation and characterisation of the p(2x1) surface of this crystal is described in detail elsewhere [3].

Thermal desorption experiments were carried out in the sample analysis chamber using a quadrupole mass spectrometer (VG). Dosing of formic acid was performed using a tube source. The sample was orientated line of sight to the quadrupole and a linear heating rate of ~ 3 Ks^{-1} applied resistively whilst masses 46, 44 and 2 were detected simultaneously.

A detailed description of the cell used for the high pressure experiments is given elsewhere[4]. Under reaction conditions a vacuum of $< 2 \times 10^{-9}$ torr could be retained in the U.H.V system whilst the cell was pressurised to ~ 4 bar. Once the sample was locked into the cell formic acid was admitted from a reservoir until the pressure, as read by a baratron, stabilised indicating the filling of the system to the partial pressure of HCOOH. This pressure was then made up to ~ 4 bar via the admission of nitrogen. The gas mixture was then recycled using a compressor. Gas sampling was made via a six way sampling valve connected to a G.C. After initiation of recycling the gas mixture, thorough mixing of the constituents was produced by recycling the gases until constant formic acid and N_2 readings were achieved. After this time the sample was then heated resistively to attain a given reaction temperature, and sampling of the gas feed made every 10 minutes until the experiment was ended. Upon completion of the experiment, the sample heating was switched off and the cell evacuated by a rotary pump. The high pressure cell volume was then pumped for ~ 30 minutes after which the sample was then retracted into the analysis chamber.

During both entry and removal of the sample from the cell area the pressure of the analysis chamber would rise to ~5×10^{-8} torr transiently, though it would normally recover to around $2-3 \times 10^{-9}$ torr within ten minutes.

3 RESULTS AND DISCUSSION

Fig 2 shows a TPD spectrum obtained for masses CO_2 and H_2 after adsorption of HCOOH to the p(2x1) surface at ~323K. H_2 evolves simultaneously with the CO_2 as well as trace amounts (~1/45th of CO_2 peak intensity) of molecular formic acid (not shown). Under the conditions applied T_{max} for the desorption occurs at 445K. The overall reaction mechanism follows that of Cu{110}, that is, formation of a formate intermediate followed

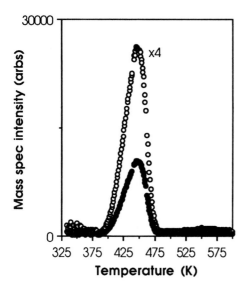

Fig 2. Obtained TDS spectra from formate adsorbed on the CuPd[85:15]{110} p(2x1) surface.
Open circles = CO_2 Filled circles = H_2.

by a thermally induced dehydrogenation leading to desorption of CO_2 and H_2. The acid proton that dissociates from the formic acid molecule to facilitate formation of the formate is known to desorb around room temperature, outside of the temperature range of these TDS experiments. This behaviour is very different from the observations on Pd {110} [5] where the rate of dehydrogenation is much faster (T_{max}~237K). An alternative dehydration route is also observed.

Complete analysis of the spectra leads to a decomposition energy (E_d) for the formate of 125 ($\pm$ 3) kjmol^{-1} and a pre exponential factor (v) of ~1 x 10^{14} s^{-1}. Madix et al [6,7] determined the parameters for the same experiment on Cu {110}, and found E_d = 136.4($\pm$3) kJmol^{-1} and v = 9.4 x 10^{13} s^{-1}. Clearly therefore the presence of the ordered underlayer of Pd perturbs the reactivity of the surface such that the alloy is now more reactive toward formate dehydrogenation than Cu{110} whilst retaining the same overall mechanism of reaction.

Fig 3 shows some typical results from the reaction carried out over the CuPd[85:15]{110} p(2x1) surface at high pressure. The traces derived from computer integration of the detected GC peaks show the consumption of HCOOH during the experiments together with the evolution of CO_2 (H_2 is not shown for clarity). The intensity of the peak derived from the carrier gas (N_2) is also displayed. These peaks were the only features seen in the GC spectra. The small irregularities in the product peaks can be clearly reconciled with the small variations in the observed intensity of the carrier gas levels and as such are artefacts of the reactor system.

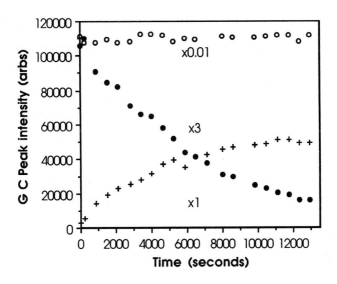

Fig3.Kinetics of the decomposition of HCOOH over CuPd [85:15] {110}p(2x1) at 483K. Filled circles: HCOOH, +: CO_2, Open circles; N_2

From the first order rate expression {1},

$$[HCOOH] = [HCOOH]_o\, e^{-kt} \quad \{1\}$$

A plot of Ln of the intensity of the HCOOH peak versus time leads to a straight line, indicating a first order dependance of the reaction rate on $HCOOH_g$.

This plot leads to a value for the rate constant k from the gradient of the line obtained. The first order nature of the reaction is verified by plotting Ln of the instantaneous rate of reaction versus Ln[HCOOH] which gives a slope (representing the order of the reaction) of ~1($\pm$0.12) (not shown). The rate constant (k) relates to the apparent activation energy for this reaction (E_{app}) thus:

$$k = v_r\, e^{-E_{app}/RT} \quad \{2\}$$

Where R is the gas constant, T is the reaction temperature (K), and v_r is a pre-exponential factor for the process. An Arrhenius plot of Ln(k) versus 1/T leads to fig 4 (filled circles for the clean surface) for reactions carried out directly after Ar ion cleaning of the surface. The resulting activation energy

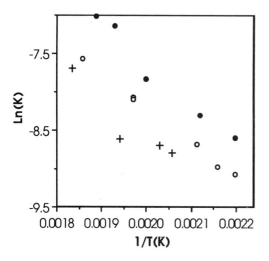

Fig 4. Arrhenius plots for the decomposition of HCOOH over sample after differing times spent in the HPC.
Filled circles; after Ar cleaning.
Open circles; +24 hrs, Crosses; +48 hrs.

derived for the overall reaction is 36 ($\pm$6) kJmol^{-1}.

It is obvious that despite the formate species being the most stable at the surface under these conditions, the apparent activation energy that we measure for the high pressure formic acid decomposition is far less than that for the formate dehydrogenation experiment under UHV. Thus the latter reaction is not solely responsible for controlling the rate of reaction of formic acid.

The simplest form of the kinetics for this system is that of a pre-equilibrium scheme viz.

$$[A] \underset{-1}{\overset{1}{\rightleftharpoons}} [I] \quad \{3\}$$

$$[I] \overset{2}{\longrightarrow} [B] + [C] \quad \{4\}$$

B and C being products (in this case CO_2 and H_2), A being the reactant species (HCOOH), and [I] being an adsorbed intermediate. Thus far the identity of this intermediate is not definitely identified at high pressure, but the results of TPD experiments would indicate that it is likely to be the adsorbed fomate species (even though the temperature range of the high pressure experiments is above that of the formate decomposition peak temperature under U.H.V conditions).

The general form of the rate expression for the above reaction is

$$d[B]/dt = k[A] \quad \{5\}$$

where k is the complex rate constant given by

$$k = k_1k_2/k_{-1} = (A_1A_2/A_{-1}) \exp^{(-(E_{app})/RT)} \quad \{6\}$$

with $E_{app} = E_1 + E_2 - E_{-1}$

There are a range of assumptions used in this model, perhaps the most important being a steady state low coverage of I (the formate) and a fast equilibrium in step 1. The latter reaction is the reverse reaction of formate hydrogenation which in this scheme is rapid. That this reaction takes place is without doubt, since formic acid desorption is seen concurrent with CO_2 and H_2 evolution in TPD experiments on Cu{110}[6,7] and CuPd[85:15]{110}p(2x1). As formate formation from formic acid in U.H.V appears facile [3,6,7] then there is no net barrier to adsorption and $E_1 \rightarrow 0$, thus

$$E_{app} \cong E_2 - E_{-1} \quad \{7\}$$

E_2 has been determined to be ~125 kJmol^{-1} from the TPD measurements. As E_{app} is determined as ~ 36 kJmol^{-1} this means that E_{-1} is of the order of 90 kJmol^{-1}. This corresponds to the energy required to take the formate and an adsorbed hydrogen into the gas phase as formic acid. Since there is no net barrier for the gas phase reverse reaction, 90 kJmol^{-1} represents the enthalpy difference between the adlayer and gas phase formic acid.

These results indicate a somewhat lower value for E_{app} for this reaction than had previously been obtained from experiments on multi crystalline CuPd alloys (for a review see [8]). These experiments indicate apparent decomposition energies of ~90 kJmol^{-1} for Cu. In the compositional range of the alloy we are studying the reported decomposition energies are of the order of 70 - 85 kJmol^{-1} depending on the pre-treatment of the alloys. Within this compositional range, tempering of the samples to produce the ordered alloys results in lower decomposition energies than is derived from disordered samples. There are a number of possible reasons for this, perhaps the most likely being the poly crystaline nature of the samples themselves. Though not detailed in [8], these will more than likely be dominated by {111} and {100} surfaces. It is well known that the rates of some reactions can vary by orders of magnitude from one crystal plane to another. It is therefore possible that our low activation energy is derived from a plane dependence in the reaction scheme whereby the reaction is far more facile on the open {110} plane than on the more densely packed, lower surface free energy {111} or {100} planes

Further experiments were performed that involved reacting the surface as before, but instead of removing the crystal from the cell, the sample was left in the cell overnight under the vacuum derived from the rotary pumping of the high pressure cell assembly. These surfaces would then be reacted as previously. The Arrhenius plots for these reactions are also shown in fig 3 (open circles and crosses). These give apparent activation energies of ~30 kJmol^{-1} though the absolute rates of reactions decrease significantly with the time that the sample has spent in the H.P.C. Thus keeping the sample under a moderate vacuum seems, unsurprisingly, to lead to a contamination of the surface that poisons the decomposition reaction without significantly affecting the apparent activation .

We will go on in the near future to compare these results with Cu{110} by carrying out high pressure measurements on that metal surface.

CONCLUSIONS

1. The CuPd[85:15]{110}p(2x1) surface shows increased activity for the dehydrogenation of adsorbed formate with respect to the Cu{110} surface. The deduced structure for this surface would indicate that this is due to an electronic perturbation of the surface due to the Pd in the second layer of the selvedge.

2. Studies of formic acid decomposition at this surface reveal a low E_{app} for this system. Invocation of a pre-equilibrium between $HCOOH_g$ and an adsorbed formate species indicates that E_{app} is the difference between the decomposition energy of the formate and that of the hydrogenation of the formate to $HCOOH_g$.

3. A poisoning of the HCOOH decomposition reaction is noted for reactions performed after the sample has spent time in a poor vacuum. This effect reduces the rate of reaction significantly but has little effect on E_{app}. The cause of this poisoning is not yet determined.

ACKNOWLEDGEMENTS

We would like to thank ICI plc and SERC for the funding of the studentship to MN.

REFERENCES

1) See for instance, G C Chinchen, P J Denny, J R Jennings, M S Spencer, and K C Waugh; Applied Catalysis, 36 (1988) 1.
2) D J Holmes, D A King, C J Barnes; Surf Sci 227 (1990) 179.
3) M A Newton, S M Francis, Y Li, D Law, M Bowker; Surf Sci 259 (1991) 45.
4) M Bowker, S M Francis, M A Newton, W G Caine, G Bromerley; To be published.
5) N Aas, Y Li, M Bowker; Proc ISSC-9 Ed. B Hayden, J Phys C Condensed Matter, 3 (1991) S281.
6) D H S Ying, R J Madix; J Catal 61 (1980) 48.
7) M Bowker, R J Madix; Surf Sci 102 (1981) 542.
8) P Mars, J J Scholten, and P Zwietering; Adv Catal 14 (1963) 35.

Enantioselective Hydrogenation: Use of Low Surface Coverages of Cinchonidine Modifier to Determine Site Character in the Platinum-catalysed Conversion of Pyruvate to *R*-(+)-Lactate

K. E. Simons[1], A. Ibbotson[2], and P. B. Wells[1]

[1] SCHOOL OF CHEMISTRY, THE UNIVERSITY, HULL, HU6 7RX, UK
[2] ICI FINE CHEMICALS MANUFACTURING ORGANISATION, PO BOX 42, HEXAGON HOUSE, BLACKELY, MANCHESTER, M9 3DA, UK

1 ABSTRACT

Optical yields and reaction rates are reported for the enantioselective hydrogenation of methyl pyruvate at 293 K and 10 bar pressure catalysed by cinchonidine-modified Pt/silica (EUROPT-1) at low alkaloid loadings (0.01 to 1.00 mg (100 mg catalyst)$^{-1}$). The results are interpreted in terms of a model which permits enantioselective hydrogenation to occur both at A-type sites involving the cooperative effect of several alkaloid molecules in a non-close-packed array, and at B-type sites at the edges of Pt particles involving a 1:1 interaction between alkaloid and pyruvate. A qualitative analysis of the optical yield versus loading curve indicates that both A- and B-type sites exist on the modified EUROPT-1 surface, whereas experiments reported by Garland and Blaser seem to be consistent with the existence of mostly B-type sites on a Pt/alumina modified by dihydrocinchonidine.

2 INTRODUCTION

The achievement of metal-catalysed enantioselectivity requires that reaction takes place in a chiral environment. The present investigation is devoted to the development of a model that describes the chiral environment obtained when members of the cinchona family of alkaloids are adsorbed onto Pt surfaces and these surfaces are subsequently used as catalysts for alpha-ketoester hydrogenation at room temperature or thereabouts.

In previous papers we have developed a mechanism for the reaction first reported by Orito and co-workers[1] in which methyl pyruvate, MeCOCOOMe, is hydrogenated to methyl lactate, MeCH(OH)COOMe, by supported Pt catalysts. Catalysts used conventionally give racemic methyl lactate; however pre-adsorption of, say, cinchonidine onto the Pt surface directs reaction (Figure 1) so that R-(+)-methyl

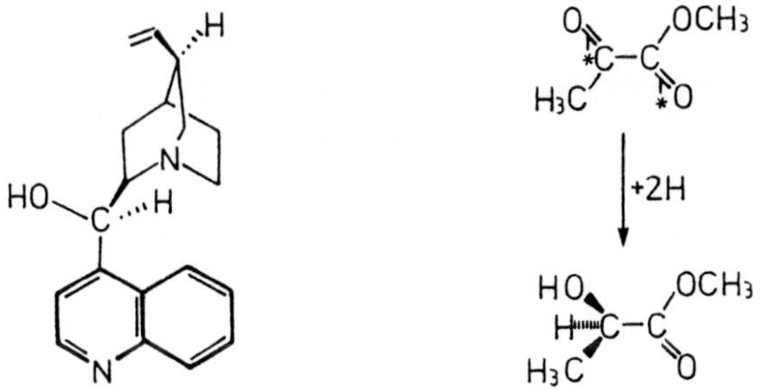

Figure 1 The alkaloid cinchonidine (left) and the overall process of R-(+)-methyl lactate formation over cinchonidine-modified Pt (right).

lactate is formed in high optical yield (60 to 90%) whereas pre-adsorption of the near diastereoisomer cinchonine provides S-(-)-methyl lactate with similar selectivity. Contary to expectations, enantioselective hydrogenation occurs at a greatly enhanced rate. Our mechanism[2] is based on the hypothesis that the L-shaped alkaloid molecules form a non-close-packed ordered array on the Pt surface, leaving exposed shaped ensembles of Pt atoms. The asymmetric hydrogenation site is then defined as this shaped ensemble of Pt atoms together with that part of the environment provided by the alkaloid molecules which bound the ensemble. This mechanism interprets both the sense of the enantioselectivity observed when cinchonidine or cinchonine are used as modifiers,[2] and the greatly enhanced rate.[3] Moreover, it accounts well for both the enantioselectivities and rates obtained when a variety of substituted or derivatised cinchona alkaloids are used as modifiers.[4]

Figure 2 is a representation of the platinum surface modified by cinchonidine.[4] The asymmetric hydrogenation site A is accessible to methyl pyruvate provided adsorption is in the configuration required to give R-(+)-lactate on hydrogenation. The spatial relationship of adsorbed methyl pyruvate to molecule X of cinchonidine is such as to permit the hydrogen bonding interaction (necessary for the enhancement of rate) that is proposed to occur when the reactant is converted to its half-hydrogenated state. Cinchonidine molecule Y directly prevents methyl pyruvate adsorbing in a form that would give S-(-)-lactate as product, and molecules Z1 and Z2 form the remainder of the boundary of the asymmetric hydrogenation site. Such boundary molecules are obviously not required at the edges of the Pt crystallites (site B). Site C represents an area of unmodified surface where slow racemic hydrogenation may proceed.

Figure 2 represents an oversimpification of the real situation in that, although high enantioselectivities are obtained, nevertheless the rate of S-(-)-lactate formation over cinchonidine-modified Pt is substantially enhanced

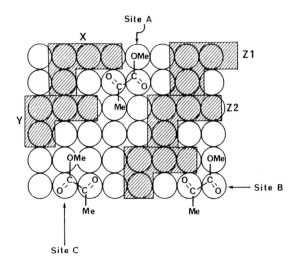

Figure 2 A representation of (a) a non-closepacked array of L-shaped cinchonidine molecules onto a Pt surface, and (b) sites A, B and C (see text).

over that observed in the racemic reaction. Thus, the ordered adsorption of the alkaloid molecules is probably less regular than that shown in Figure 2.

The model in Figure 2 represents a total of four alkaloid molecules as contributing to asymmetric hydrogenation site A, although only two play a crucial role; one alkaloid molecule is sufficient at site B. Remembering that in each case a 1:1 pyruvate/alkaloid interaction is responsible for the enhanced rate, we now look for experimental evidence which informs as to the total number of alkaloid molecules which define the asymmetric hydrogenation site in the average case.

Our approach has been to examine optical yields and reaction rates for reactions in ethanolic solution involving very small concentrations of modifier. Results of a parallel study involving reactions in toluene solution have been reported.[5]

3 EXPERIMENTAL

The reference catalyst EUROPT-1, a 6.3% Pt/silica manufactured by Johnson Matthey and characterised by the EUROCAT group[6,7] was used in all experiments. The Pt active phase is substantially oxidised in the as-received material.[8] Catalyst samples were used either as-received (State 1 as defined in reference 2) or after reduction at 373 K in 1 bar hydrogen for 0.5 h (similar to State 3[2]). Methyl pyruvate (Fluka), cinchonidine (Aldrich), ethanol and hydrogen were used without purification.

The Fischer Porter high pressure stirred reactor with computer controlled hydrogen addition facility has been described.[2]

The preparation of an enantioselective catalyst is achieved by the process of 'modification' in which a chiral substance (the modifier) is adsorbed onto the catalyst surface from the solution at ambient temperature. In this work the mass of the cinchonidine modifier present in 20 ml solvent was varied from 0.01 to 1.00 mg and 100 mg of fresh EUROPT-1 was used in each experiment.

Two series of experiments were carried out. In Series I, 100 mg of as-received EUROPT-1 was placed in the reactor, 20 ml of modifier solution in ethanol was admitted, and the mixture was stirred in air for 1 h at 293 K. 10 ml of methyl pyruvate were then added. The vessel was flushed first with nitrogen and then with hydrogen, and hydrogen pressure increased to 10 bar. The pressure was maintained at this value by computer control and hydrogen uptake was monitored as a function of time.

In Series II, 100 mg EUROPT-1 was reduced in 760 Torr hydrogen in a separate vessel and the cinchonidine solution in ethanol was admitted before removal of the hydrogen. Catalyst and modifier solution were then transferred to the reactor, and the mixture was stirred in air for 1 h at 293 K. Air was removed and the reactants added as described for Series I experiments.

The step involving stirring the catalyst with the modifier solution in air was incorporated because it forms part of the original published procedure.[1] We have demonstrated (i) that the time required for this treatment may be reduced from 18 h (Orito's recommendation) to 1 h without reducing the optical yield but with some loss of reaction rate, and (ii) that omission of this step under normal conditions (i.e. anaerobic modification) results in a reduced optical yield and a substantial loss of activity.[9] In the present work with reduced modifier concentrations this step was sometimes omitted and the effects are recorded.

The addition of ethanol or of ethanolic solutions of cinchonidine to reduced EUROPT-1 resulted in visible hydrogen evolution caused by the dissociative adsorption of ethanol. We presume that under reaction conditions at 10 bar hydrogen pressure this reaction was at least partially reversed. This phenomenon is the subject of a separate investigation.

Reactions were stopped when the desired conversion had occurred: the procedures for separation by rotary evaporation, analysis by glc, and determination of optical yield by polarimetry have been described.[2] Optical yield is defined as (%R-(+)-methyl lactate - %S-(-)-methyl lactate).

4 RESULTS

The conditions originally recommended by Orito and coworkers,[1] and used as standard in our previous work[2,3,9] involved catalyst modification in a solution containing 200 mg cinchonidine in 40 ml ethanol. Solution containing excess alkaloid was decanted off, and moist catalyst transferred to the reactor where 20 ml fresh ethanol and 10 ml pyruvate were added before pressurisation with hydrogen. Under these conditions optical yields of 70 ±5% were routinely obtained at conversions of 90% or above. Repetition of this procedure using 11 mg and 1 mg cinchonidine gave optical yields of 49% (at 47% conversion) and 30% (at 26% conversion) respectively. It was clear, therefore, that experiments could be conducted in an interesting region where there was insufficient modifier properly to template the surface. Accordingly the procedure described in the Experimental Section was developed in which all of the modifier solution is maintained in contact with the catalyst.

The variation of optical yield in methyl pyruvate hydrogenation with the weight of cinchonidine added to the catalyst is shown in Figure 3. Reactions provide an excess of (R)-(+)-methyl lactate. Optical yields reach a plateau value of about 66% (i.e. 83% (R)-(+)-, 17% (S)-(-)-) as the amount of added cinchonidine approaches about 0.70 mg per 100 mg catalyst. Results of experiments represented by open points (Series I) and by filled points (Series II), fall on the same curve, indicating that the alkaloid is

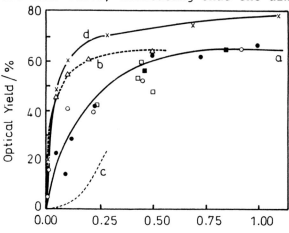

Figure 3 Dependence of optical yield on mass of alkaloid present in the reactor. Curve (a), experimental work; curves (b) and (c), theoretical (see text); curve (d) experimental dependence for an analogous system taken from reference 13.

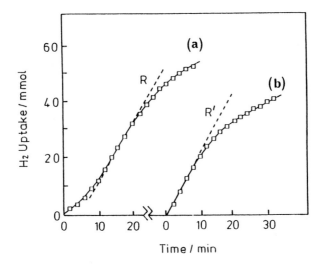

Figure 4 Definition of reaction rates R and R'

equally effective whether it is adsorbed on the catalyst initially in either the oxidised or reduced state. Similar optical yields were observed irrespective of whether the catalyst was stirred in air with the modifier (circles), or not (squares). When toluene is used as a solvent the points fall on the same curve; these results have been published in reference 5.

Reactions carried out using the high concentrations of modifier recommended by Orito exhibit uptake versus time curves as shown in curve (a) of Figure 4 in which hydrogenation at constant rate, R, is preceded by a period of acceleration. Such uptake versus time curves were obtained in the present work when the cinchonidine loading was in the range 0.7 to 1.0 mg per 100 mg catalyst. When the loading was 0.5 mg and the catalyst had been stirred in the modifier solution in the presence of air, the acceleratory period was absent. For loadings below 0.5 mg the acceleration period was short or absent under all conditions. When there was no acceleration the initial rate R', was measured as shown in curve (b) of Figure 4.

The variation of R or R' with the weight of cinchonidine added to 100 mg of catalyst is shown in Figure 5. Rate increased as the loading of alkaloid increased. Although there is experimental scatter, no plateau is observed. Rates of reaction in toluene solution were faster than those observed in ethanol[5] (as has been reported by Blaser, Baiker and coworkers[10]) and a plateau was observed. The rate of reaction using a catalyst loaded with 1 mg of modifier exceeded that of an unmodified catalyst by a factor of over 20. This value concurs with that reported in reference 2 for EUROPT-1 modified at the normal high concentrations recommended by Orito et al.

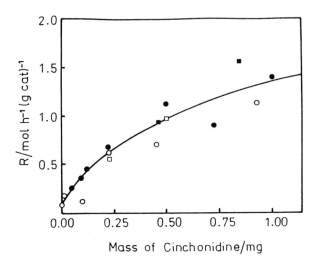

Figure 5 Dependence of rates, R or R' on mass of alkaloid present in the reactor. Weight of EUROPT-1 100 mg.

5 DISCUSSION

The hypothesis that cinchonidine and related alkaloids template Pt surfaces by the formation of non-close-packed ordered adlayers has contributed greatly to our present understanding of the asymmetric hydrogenation site,[4] and evidence for such ordered adsorption on Pt single crystal surfaces is currently being sought.[11] However, Figure 2 shows that fully structured sites represented by site A are likely to be accompanied by 1:1-interaction sites represented by site B, the latter being located at edges of the Pt crystallites. Since supported catalysts contain small metal particles (typically 1.8 to 2.0 nm in diameter for EUROPT-1[7]) the contribution of B-type sites may be considerable. Examination of the behaviour of (a) a single modifier at low concentrations, and (b) mixtures of modifiers at high concentration, provide two independent methods for the evaluation of the average number of alkaloid molecules that contribute to the architecture of asymmetric hydrogenation sites.

Results obtained by use of mixtures of modifiers is described elsewhere.[12] When EUROPT-1 was modified by a mixture of two alkaloids, one of which, (when used on its own) favoured R-(+)-lactate, and the other S-(-)-lactate, the resulting optical yield was to a first approximation that expected on the basis of an additive effect of the components. On the assumption that reaction occurred at A-type sites, it was concluded that molecules Z1 and Z2 (Figure 2) may either be chemically identical to X and Y (as previously postulated) or they may have the near mirror image configuration. Thus, ordered adsorption as originally envisaged[2,4] may be an unnecessarily restrictive prerequisite for the formation of site A. The contribution of B-type sites was not discussed.

Preliminary results for the effect of sub-monolayer concentrations of alkaloid on optical yield and reaction rate for reactions in toluene solution have been published.[5] Under these conditions, where the majority of the alkaloid is adsorbed from solution onto the catalyst, the minimum quantity of alkaloid required to achieve the highest attainable optical yield over EUROPT-1 corresponded closely with that required by the template hypothesis to populate the Pt surface. The present work extends and refines that study by examination of the rather different surface conditions that prevail in the same system when ethanol is used as solvent.

The adsorption of even very small quantities of cinchonidine onto EUROPT-1 results in enhanced rate (Figure 5) and enantioselective reaction (Figure 3). The optical yields recorded for reactions in ethanolic solution concur with those reported for toluene solution, although the rates were higher in toluene.[5] Optical yield reaches a plateau value at a loading of about 0.70 mg cinchonidine per 100 mg catalyst, and this value of 65% is the same as that observed for State 1 EUROPT-1 under standard conditions involving high loadings of modifier.[2] However, rates R or R' do not reach a plateau (unlike the situation for reactions in toluene solution) nor do they achieve the value of 1.65 mol h^{-1} (g cat)$^{-1}$ typical of standard conditions with higher modifier loadings. These characteristics demonstrate that the surface does not become saturated with cinchonidine within the loading range here studied, and this is consistent with the surface being partially covered by the product of the dissociative adsorption of ethanol.

The manner in which optical yield rises with alkaloid loading is diagnostic as to the average number of alkaloid molecules that are associated with each asymmetric hydrogenation site. If optical yield, like the enhanced rate, was purely the result of a 1:1 interaction of alkaloid and reactant, then optical yield should rise very steeply with loading. By use of the rates recorded for racemic and enantioselective reaction (Figure 5) and an assumed optical yield of 65% for the enantioselective component of the total reaction, the variation of optical yield with loading for the 1:1 interaction model is calculated to be as shown in curve (b) of Figure 3. On the other hand, if an ordered array of four alkaloid molecules was the only configuration capable of giving enantioselective reaction, and adsorption occurred without cooperative effects (e.g. leading to island formation), then an induction period should be observed in the attainment of optical yield as shown in curve (c) of

Figure 3. The experimental results show an intermediate situation, but with no induction period.

At this point, we note the published work of Garland and Blaser for the hydrogenation of ethyl pyruvate in ethanol solution at 293 K and 20 bar pressure over Pt/alumina modified by dihydrocinchonidine;[13] their reported variation of optical yield with loading is shown as curve (d) in Figure 3. This behaviour is clearly interpretable in terms of a 1:1 interaction model.

We conclude that, over cinchonidine-modified EUROPT-1, the average asymmetric hydrogenation site involves the participation of more than one alkaloid molecule. At the lowest alkaloid loadings, B-type sites and 1:1 interactions will predominate unless cooperative effects in alkaloid adsorption are important. At higher loadings, both A- and B-type sites contribute, causing the experimental points to fall below the curve (b). It should be recalled that the Pt particles in EUROPT-1 appear to posses raft-like character,[14] so the Pt atom array shown in Figure 2 is particularly appropriate. Thus, supposing that a fully templated Pt surface might provide equal numbers of A- and B-type sites, and remembering that the adlayer contains also the products of ethanol dissociative adsorption (which could serve in place of alkaloid molecules Z1 and Z2 to bound the site) the system should behave as though the average alkaloid:pyruvate ratio is in the range of 1.5 to 2.5. This analysis confirms the general applicability of the model shown in Figure 2; it provides evidence for the very considerable importance of 1:1 interactions and the B-type sites, and suggests that the products of ethanol adsorption play a role in the definition of the asymmetric hydrogenation site.

These developments of the model indicate that important information can now be obtained from studies of the effects of platinum particle morphology on optical yield, particularly at low alkaloid loadings. Moreover a quantitative treatment of optical yield versus loading is being developed and tested.[15] Catalysts containing very small particles, or larger particles that are highly faceted can be expected to contain a particularly high proportion of B-type sites, whereas catalysts having large Pt particles containing extensive low index planes should contain a high proporton of A-type sites. The Pt/alumina studied by Garland and Blaser[13] was of low dispersion (D=0.28) but behaved as though it contained largely B-type sites; we speculate that the Pt particles in this catalyst were highly faceted.

6 ACKNOWLEDGEMENTS

We thank SERC for a CASE studentship (K.E.S) for which the cooperating body was ICI Fine Chemicals Manufacturing Organisation.

REFERENCES

1. Y. Orito, S. Imai and S. Niwa, Nippon Kagaku Kaishi, 1979, 1118; 1980, 670; 1982, 137.
2. I.M. Sutherland, A. Ibbotson, R.B. Moyes and P.B. Wells, J. Catal., 1990, 125, 77.
3. G. Bond, P.A. Meheux, A. Ibbotson and P.B. Wells, Catal. Today, 1991, 10, 371.
4. G. Webb and P.B. Wells, Catal. Today, 1992, 12, 319.
5. G. Bond, K.E. Simons, A. Ibbotson, P.B. Wells and D.A. Whan, Catal. Today, 1992, 12, 421.
6. G.C. Bond and P.B. Wells, Appl. Catal., 1985, 18, 225.
7. J.W. Geus and P.B. Wells, Appl. Catal., 1985, 18, 231.
8. R.W. Joyner, J. Chem. Soc. Faraday Trans. I, 1980, 76, 357.
9. P.A. Meheux, A. Ibbotson and P.B. Wells, J. Catal., 1991, 128, 387.
10. J.T. Wehrli, A. Baiker, D.M. Monti, H.U. Blaser and H.P. Jallet, J. Mol. Catal., 1989, 57, 245.
11. R.K. Rajumon, M.W. Roberts and P.B. Wells, unpublished work.
12. K.E. Simons, P.A. Meheux, A. Ibbotson and P.B. Wells, accepted for 10th Intern. Congr. Catalysis, Budapest, July 1992.
13. M. Garland and H.U. Blaser, J. Am. Chem. Soc., 1990, 112, 7048.
14. S.D. Jackson, M.B.T. Keegan, G.D. McLellan, P.A. Meheux, R.B. Moyes, G. Webb, P.B. Wells, R. Whyman and J. Willis, 'Preparation of Catalysts V' (G. Poncelet, P.A. Jacobs, P. Grange and B. Delmon, Eds., Elsevier, Amsterdam, 1991) p.135.
15. M.S. Spencer, private communication.

An *In Situ* Study of Pyrochlore-type Catalysts for the Formation of Synethesis Gas from Methane and CO_2

A. T. Ashcroft[1], A. K. Cheetham[1,2], R. H. Jones[3], S. Natarajan[3], J. M. Thomas[3], and D. Waller[3]

[1] CHEMICAL CRYSTALLOGRAPHY LABORATORY, UNIVERSITY OF OXFORD, 9 PARKS ROAD, OXFORD, OX1 3PD, UK

[2] MATERIALS DEPARTMENT, UNIVERSITY OF CALIFORNIA, SANTA BARBARA, CALIFORNIA, CA 93106, USA

[3] DAVY FARADAY RESEARCH LABORATORY, THE ROYAL INSTITUTION OF GREAT BRITAIN, 21 ALBEMARLE STREET, LONDON W1X 4BS, UK

1 INTRODUCTION

The rare-earth iridium- and ruthenium-containing pyrochlores have previously been shown to be excellent catalysts for the partial oxidation[1] and for the dry-reforming[2] of methane to synthesis gas. It has been established[1,3] that in catalysing these reactions, the pyrochlore structure of these materials decomposes to produce supported metal particles which appear to be the active species. Here, we study the breakdown of the pyrochlore structure in more detail by use of *in situ* X-ray diffraction with mass spectrometry using a combined rotating-anode X-ray and catalytic reactor facility,[4] and by using temperature programmed reaction spectroscopy (TPRS).

2 EXPERIMENTAL

The rare-earth iridates were prepared from the appropriate stoichiometric ratios of rare-earth oxide (Aldrich Chem. Co. Ltd.) and iridium oxide (Johnson Matthey, batch 051236A), which were intimately mixed, ground, pressed into pellets, and heated in air at 1323K for 10 days, with frequent re-grinding to ensure sample homogeneity. A similar procedure was employed for the ruthenates, except that the pellets were sealed into evacuated silica capsules and heated at 1373K for 2 days.

For the *in situ* XRD, the catalyst was packed down smoothly into a porous silica frit, through which the reactant gases (BOC Ltd., 25% CH_4, 25% CO_2, 50% He) could pass. A pre-calibrated heating coil was wound around the frit, and the entire catalytic cell was mounted on a Siemens D500 powder X-ray diffractometer with a Stoe rotating anode X-ray source. The product gases were continuously monitored using a VG Micromass PC200 mass spectrometer and an AMS Model 91 gas chromatograph equipped with a thermal conductivity detector.[4]

3 RESULTS AND DISCUSSION

In Situ Experiments On $Eu_2Ir_2O_7$ and $Gd_2Ru_2O_7$.

It can be seen from Figure 1 that the europium iridium pyrochlore, $Eu_2Ir_2O_7$, decomposes under the CH_4/CO_2 gas mixture in a similar fashion to that which was observed for the CH_4/O_2 gas mixture at 873K.[3] Figure 2 shows that the resulting

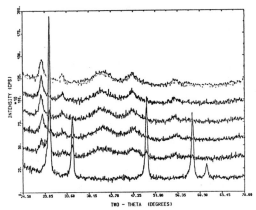

Fig. 1. X-ray diffraction patterns of $Eu_2Ir_2O_7$ under reaction conditions at 863K at time intervals of 0 (bottom), 20, 40, 60 and 80 minutes, and then at 933K.

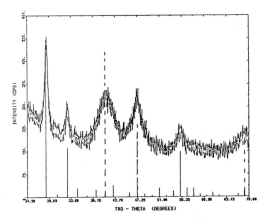

Fig. 2. X-ray diffraction pattern of the decomposed europium iridium pyrochlore at room temperature, with Eu_2O_3 (solid) and iridium metal (dashed) reflection markers.

species is, again, a mixture of iridium metal and europium oxide. The mass spectrometer trace, shown in Figure 3, shows a major difference from that which was observed for the initiation of the partial oxidation reaction.[3] In the previous work, the onset of catalytic activity was preceeded by the evolution of a burst of oxygen, which was presumed to be from the breakdown of the pyrochlore structure according to:

$$Eu_2Ir_2O_7 \rightarrow Eu_2O_3 + 2Ir + 2O_2\uparrow \quad (1)$$

In this case, there are no pronounced features in the oxygen trace, although there is, once more, a clearly defined onset of catalytic acitivity. A closer inspection of the water trace reveals a deviation similar to that observed for oxygen in the previous system, and it seems probable that the pyrochlore structure is decomposing, as it was under the partial oxidation gas mixture. Under the more reducing dry-reforming mixture, however, the oxygen may be converted to water before it can be detected. This reduction could be due to the more concentrated (50% reactants) gas

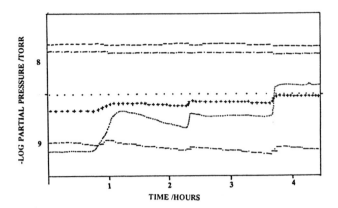

Fig. 3. Results of mass spectromeric analysis of the effluent gases from dry-reforming over $Eu_2Ir_2O_7$ as a function of time and temperature. The temperature was stepped up to 863K over the first 45 minutes, held at that temperature for 90 minutes before being heated up to 933K for 85 minutes, and then heated to 1013K. The gases are as indicated: CH_4(---), CO_2(----), CO(•••), H_2O(-•-•-), H_2(······).

mixture which it is safe to employ for this reaction, so that more hydrogen is produced than in the partial oxidation experiment (only 7.5% reactants). Alternatively, the water may be produced via the reverse water gas shift reaction:

$$CO_2 + H_2 = CO + H_2O \qquad (2)$$

The production of water is accompanied by syngas formation, but the amount of syngas produced diminishes with time. This could be due to growth of the metal particles to produce a decrease in the surface area of the active phase. The effect of raising the temperature from 863K to 933K, and again to 1013K, was to increase the production of synthesis gas as the thermodynamics became more favourable.

It can be seen from Figure 4 that the decomposition of the gadolinium ruthenium pyrochlore under the dry-reforming gas mixture is similar to that of europium iridate. However, it can also be seen from a comparison of Figures 1 and 4 that the decomposition of the ruthenate occurs more slowly, and requires a higher temperature than is necessary to provoke the decomposition of the iridate. By inspection of Figure 5, and a comparison with Figure 2, it can be seen that the products of the decomposition are, again, rare-earth oxide and metal, but in this case the metal peaks are much less clearly defined. This could be indicative of a smaller particle size for the ruthenium metal, producing a broader peak. The mass spectrometer profiles in Figure 6, once again, show increases in the amount of hydrogen, carbon monoxide and water with increased temperature. Water is presumably produced via the reverse water gas shift reaction (2), and it is possible that the dry-reforming reaction is proceeding via a combination of reverse water gas shift (2) and steam-reforming (3) reactions:

$$CH_4 + H_2O = CO + 3H_2 \qquad (3)$$

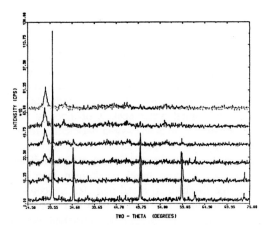

Fig. 4. X-ray diffraction patterns of $Gd_2Ru_2O_7$ under reaction conditions at 933K at time intervals of 0 (bottom) and 15 minutes, and then at 1003K at time intervals of 0, 15, 30 and 45 minutes.

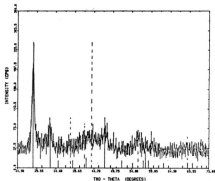

Fig. 5. X-ray diffraction pattern of the decomposed gadolinium ruthenium pyrochlore at room temperature, with Gd_2O_3 (solid) and Ru metal (dashed) reflection markers.

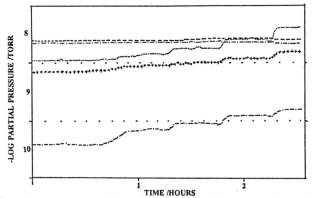

Fig. 6. Results of mass spectromeric analysis of the effluent gases from dry-reforming over $Gd_2Ru_2O_7$ as a function of time and temperature. The temperature was stepped to 933K over the first 105 minutes, held for 30 minutes before being heated to 1003K. The gases are as indicated: CH_4(---), CO_2(-·-), CO(+++), H_2O(-··-), H_2(······).

Temperature Programmed Reaction Spectroscopy (TPRS) Over $Nd_2Ir_2O_7$ and $Ln_2Ru_2O_7$ (Ln=Eu, Gd, Nd, Sm).

The temperature programmed reaction spectra of $Ln_2Ru_2O_7$ (Ln= Eu, Sm) under the methane/carbon dioxide gas mixture show broad humps in the water and oxygen traces as the catalyst is reduced, followed by a steady increase in syngas formation as the equilibrium shifts with temperature. For the neodymium and gadolinium ruthenates, Figure 7, an extra, small water peak is obtained at about 500K, which presumably accompanies the formation of an oxygen-deficient pyrochlore intermediate, as reported elsewhere.[5] The reaction spectrum from neodymium iridate, Figure 8, shows a sharper water peak before activation occurs, which indicates that either the iridium pyrochlore undergoes a more rapid breakdown than do the ruthenium-containing analogues, or that the iridium metal particles are converting the reaction mixture into a product which contains less water than that which is produced by the ruthenium metal catalyst. The latter would be consistent with previous results obtained on this system.[2]

4 CONCLUSIONS

It has been shown that the initiation of the dry-reforming reaction proceeds in a similar fashion to that previously observed for the partial oxidation reaction.[3] The main difference between the two catalytic processes is that, under the more reducing dry-reforming conditions, the lattice oxygen is reduced to water before it escapes from the reactor. The decompositions of the iridates and ruthenates appear to be different in that the latter require a higher temperature and appear, in some cases at least, to involve an oxygen-deficient intermediate.

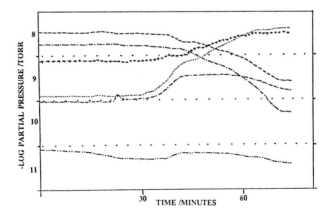

Fig. 7. Results of mass spectromeric analysis of the effluent gases from the dry-reforming TPRS over $Gd_2Ru_2O_7$ as a function of time and temperature. The temperature was increased at a rate of 10K/min from 373K to 1073K, with a data point being collected every 6s, to give a point per degree Kelvin. The gases are as indicated: CH_4(---), CO_2(-·-), CO(+++), H_2O(----), H_2(·····), O_2(-··-).

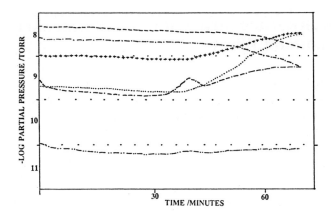

Fig. 8. Results of mass spectromeric analysis of the effluent gases from the dry-reforming TPRS over $Nd_2Ir_2O_7$ as a function of time and temperature. The temperature was increased at a rate of 10K/min from 373K to 1073K, with a data point being collected every 6s, to give a point per degree Kelvin. The gases are as indicated: CH_4(---), CO_2(-·-), CO(•••), H_2O(-··-), H_2(······), O_2(-··-).

ACKNOWLEDGEMENTS

We would like to thank the SERC, British Gas and the Royal Society for financial support, and Dr. J.W. Couves, Dr. P.D. Battle and Dr. A.M. Chippindale for encouragement.

REFERENCES

1. A.T. Ashcroft, A.K. Cheetham, J.S. Foord, M.L.H. Green, C.P. Grey, A.J. Murrell and P.D.F. Vernon, Nature, 1990, 344, 319.
2. A.T. Ashcroft, A.K. Cheetham, M.L.H. Green and P.D.F. Vernon, Nature, 1991, 352, 225.
3. R.H. Jones, A.T. Ashcroft, D. Waller, A.K. Cheetham, and J.M. Thomas, Catal. Lett., 1991, 8, 169.
4. P.J. Maddox, J. Stachurski and J.M. Thomas, Catal. Lett., 1988, 1, 191.
5. A.T. Ashcroft, A.K. Cheetham, S.M. Clark, R.H. Jones, S. Natarajan, J.M. Thomas and D. Waller, manuscript in preparation.

Chemically Stimulated Exo-electron Emission from Platinum Wires

Maureen E. Cooper
DEPARTMENT OF BIOLOGICAL AND MOLECULAR SCIENCES,
UNIVERSITY OF STIRLING, STIRLING FK9 4LA, UK

1 INTRODUCTION

Exo-electron emission is the emission of very low energy electrons (0-10 ev) from defects and traps at or near the surface of a solid.[1] It may be brought about in several ways, for example mechanical working of a piece of metal,[2] abrasion of a metal surface[3] or irradiation of a solid with ionising radiation.[4] Exo-electrons may be released after irradiation of surfaces with light from the near UV or visible range of the spectrum.[5] This phenomenon is distinct from the photo-electric effect per se since the wavelengths used are far longer than the photoelectric threshold of the material being investigated.

The emission also accompanies phase changes and many chemical reactions. For example, Lohff and Seidl have shown that exo-electron emission accompanies oxidation of zinc and copper respectively.[6,7] Sujak et al[1] measured exo-electron emission from zinc oxide as a function of temperature and found the temperature of maximum emission corresponded to the maximum catalytic activity for methanol oxidation. Ohashi et al,[8] and Sato and Seo[9] have demonstrated that the electron emission activity is proportional to catalytic activity for the partial oxidation of ethylene over silver catalysts, while Hoenig and co-workers[10] have observed exo-electron emission accompanying the catalytic oxidation of CO, H_2 and NH_3 over heated palladium wires.

2 EXPERIMENTAL

In general the low energy electrons have been detected using an electron multiplier which necessitates the use of high vacuum, or with a Geiger-Mueller counter which requires a counting gas. Both interfere with the possibility of following processes of catalytic interest directly.

Detector

The present counter comprises a modified cylindrical gas flow proportional counter (Figure 1). A 1 m long Pt wire, diameter 3×10^{-3} m was mounted on a glass former between the anode, 2.5×10^{-5} m diameter flame annealed W, and the cathode, 2.5×10^{-2} m diameter Cu. Tungsten leads provided electrical connections to a 24 V dc. supply, allowing the filament to be heated to around 900 °C. Gases were dried over anhydrous sodium carbonate and their flow rates measured by GEC-Elliot rotameters. Temperature was measured by attaching a Pt,13% Rh wire to the filament to provide the hot junction of a Pt/Pt-Rh/Pt thermocouple connected to a Derritron Potentiometer.

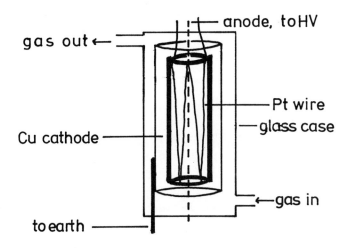

Figure 1 Modified gas-flow proportional counter containing a platinum filament which can be heated electrically to temperatures around 1000 K.

Electronics

A Nuclear Enterprises SR5 provided a 0-4 kV high voltage supply, scaler and ratemeter facilities and a linear pulse analyser. This was used with a charge-sensitive NE 5287SC preamplifier with a modified FET 5287AC input stage.

This counter design gave reliable counting characteristics in a variety of gases at temperatures up to 800 K.[11]

Experimental Procedure

Two main types of experiments were carried out. In the first, short heating experiments, the wire was heated for around 30 s in the gas under investigation, to allow the temperature to stabilise, then the wire was cooled in the same gas. Flow rates were maintained at 14.4 $dm^3\ h^{-1}$. Counts and temperature were monitored as the wire cooled and temperatures of any peak maxima noted. New gases were always introduced with the filament at room temperature.

In the second, prolonged heating experiments, the filament was heated continuously in several gases in sequence before the 'cooling curve' was obtained to observe if there was any effect due to the history of the wire.

Scanning Electron Microscopy

Sections of used and unused platinum were examined using a Philips PSEM 500 Scanning Electron Microscope to determine whether the surface had altered in any way and to look for signs of carbide formation.

3 RESULTS

Short-heating Experiments

Some typical results are shown below; see reference 11 for more detail.

<u>Ar/10% methane</u> Several different cooling curves were observed when Pt cooled in this gas mixture (Figure 2), the detailed structure depending on the preceding gas.

<u>Ar/10% ethane</u> Although the first run in Ar/ethane gave almost no emission, it increased rapidly with each subsequent run (Figure3).

<u>Ar/10% ethylene</u> Emission maxima at around 55 °C were observed in the cooling curve for Ar/ethylene preceded by Ar/methane or Ar/acetylene. When preceded by Ar/methane there was also a large peak at 46 °C.

<u>Ar/10% acetylene</u> In general there was almost no emission except when this run was preceded by Ar/methane, when two peaks were observed at 77 °C and 51 °C.

<u>Nitrogen</u> Emission was only observed from clean Pt filaments or after treatment in oxygen.

<u>Oxygen</u> Emission fell rapidly to zero from around 7000 cps. Consecutive cycles in oxygen gave no subsequent emission except where the Pt had been exposed to acetylene.

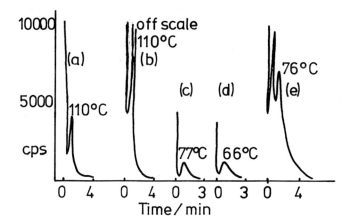

Figure 2 cps v time for a Pt filament cooling in Ar/10% methane. The preceding runs were in (a) N_2, H_2, Ar, O_2, or Ar/methane; (b) acetylene; (c) Ar/acetylene; (d) Ar/ethylene; (e) Ar/ethane.

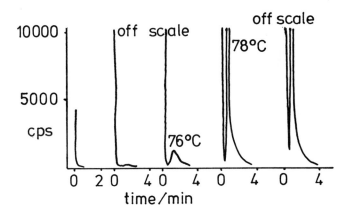

Figure 3 cps v time for a platinum filament cooling in Ar/10% ethane in five consecutive runs from left to right.

Long-heating Experiments

Here the wire was heated in one gas then several others were introduced in sequence, allowing the count-rate to settle after each introduction. Finally the filament was allowed to cool. Typical results are shown below.

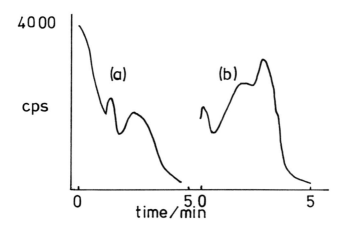

Figure 4 cps v time for Pt filament cooling in Ar/methane after heating (a) in O_2 (N_2 flush); (b) in H_2.

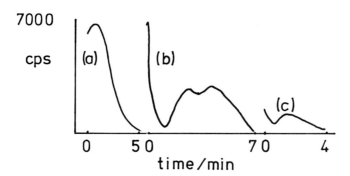

Figure 5 cps v time for a Pt filament cooling in N_2 after long heating experiments. The wire was previously heated in (a) O_2; (b) Ar/methane; (c) Ar or He.

Ar/10% methane Once again varied cooling curves were observed (Figure 4).

Nitrogen There were also varied cooling curves for a Pt wire cooling in N_2, with significant emission after oxygen treatment (Figure 5).

Oxygen O_2 showed much higher levels of emission than with the short heating experiments.

4 DISCUSSION

Observation of a series of common temperatures for emission maxima suggests the presence of common surface species. This is reinforced by the fact that some of the temperatures are only found after pretreatment of the Pt with saturated hydrocarbons. Others are common to both saturated and unsaturated hydrocarbons. In addition, extensive surface rearrangement was evident from the micrographs of the Pt wires before and after treatment.[11] A 'new' wire showed extensive groove formation from the extrusion process. An 'old' wire showed grain growth, with considerable roughening, indicative of the early stages of faceting. Prolonged heating in oxygen alone had little effect on the surface morphology, indicating that hydrocarbons must have a considerable influence on the stability of some Pt crystal planes while other surfaces remained relatively unchanged.

From the literature it seems reasonable to suggest that exo-electron emission accompanies the making or breaking of bonds at the surface, so that it should be seen to accompany adsorption and desorption of gases on a Pt surface, or rearrangement or relaxation of the Pt surface atoms.

REFERENCES

1. B. Sujak, T. Gorecki and L. Biernacki, Acta Phys Pol, 1971, A39, 147.
2. A.H. Meleka and W.Barr, Nature, 1960, 187, 232.
3 J. Kramer, 'Der Metallische Zustand', Vandenhoeck & Ruprecht, Goettingen, 1950.
4. H. Hieslmar, H. Mueller, Z Phys, 1958, 152, 642.
5. J. Ramsey and G.F.J. Garlick, Brit J Appl Phys, 1964, 15.1353.
6. J. Lohff, Z Phys, 1956, 145, 504.
7. R.Seidl, Acta Phys Aust, 1957, 10, 402
8. H. Ohashi, H. Kano, N. Sato & G. Okamoto, J Chem Soc Japan Ind Chem Sect, 1966, 69, 997.
9. N. Sato and M. Seo, J. Cat, 1972 24, 224.

10. S.A. Hoenig and J. R. Lane, Surface Science, 1968, 11, 163.
11. M. Cooper and S.J. Thomson, JCS Faraday 1, 1980, 76 1923.

Surface Characterization of Pt/TiO$_2$ Catalyst by HRTEM and FTIR Investigation. H$_2$ and CO Adsorption on a Pre-oxidized Sample

L. Marchese, G. Martra, S. Coluccia, G. Ghiotti, and A. Zecchina

DIPARTIMENTO DI CHIMICA INORGANICA, CHIMICA FISICA E CHIMICA DEI MATERIALI DELL'UNIVERSITÀ DI TORINO, VIA P. GIURIA 7, 10125 TORINO, ITALY

ABSTRACT

Oxidized platinum produced by O$_2$ treatment at 773 K on a Pt/TiO$_2$ catalyst is reduced to Pt^0 by hydrogen at 300 K. Hydrogen dissociates on the newly formed Pt^0 sites, producing hydrides which absorb in the 2150-2030 cm^{-1} range. Reduction of the TiO$_2$ support occurs by hydrogen spillover from the metal phase. This causes severe loss of transparency due to the scattering of IR light by free carriers. Spectra of adsorbed CO confirm that hydrogen treatment at 300 K reduces exposed Pt$^{\delta+}$ to Pt^0.

1 INTRODUCTION

The importance of reforming reactions has encouraged studies on surface characterization of platinum catalysts. Growing attention has been attracted by Pt/TiO$_2$, due to the interest for possible promoting effects of the reducible support[1]. CO and H$_2$ adsorption have been often adopted for characterizing the Pt metal phase produced on the catalysts upon standard pretreatments which generally require final reduction in H$_2$ at 773 K.

Important results have been obtained by this approach for Pt/Al$_2$O$_3$ and Pt/SiO$_2$. However, H$_2$ at such high temperature in the case of Pt/TiO$_2$ produces significant reduction of the support[1,2] which causes a severe loss of transparency in the whole infrared region[3]. Large SMSI (Strong Metal-Support Interaction) effects are produced at the same time[1,2]. To observe the stretching modes of adsorbed CO species, it is necessary to admit oxygen onto the sample[3]. This treatment partly restores the transmittance, but certainly modifies the surface under examination. Due to these difficulties, studying the evolution of the surface at the various steps of the pretreatment, and particularly at various stages of reduction, may help in characterizing Pt/TiO$_2$ samples.

Comparison of the effects of reduction at high (773 K) and low (≤523 K) temperature gave information on the building up of SMSI.[1,2] The results of hydrogen treatments at room temperature are described here, showing that Pt surface metal phase is produced in such mild conditions, together with reversible reduction of the support.

2 EXPERIMENTAL

Pt/TiO_2 catalysts (2.8 wt% Pt) were provided by the Laboratoire de Réactivité de Surface et Structure - Université P. et M. Curie - Paris. The samples were prepared by impregnation of TiO_2 (Degussa P25) with hexachloroplatinic acid solutions and dried at 393 K. Self supporting pellets for IR measurements, following standard pretreatment, were calcined at 773 K 1 hour in 13.3 kNm^{-2} O_2 (1 Torr = 133 Nm^{-2}), outgassed (5 min) at 773 K, reduced in H_2 (13.3 kNm^{-2}) 1 hour at 773 K and finally evacuated for 5 min at the same temperature. Prior to the measurements described in this report, the sample was re-oxidized (13.3 kNm^{-2} O_2, 1 hour) at 773 K and outgassed at the same temperature 5 min.

The IR cells allowed all thermal treatments and adsorption-desorption experiments to be carried out in situ. The spectrometer was a Bruker IFS 48; the resolution was 4 cm^{-1}. The HRTEM microscope was a Jeol 2000 EX. High purity gases (UCAR) were used with no further purification except liquid nitrogen trapping.

3 RESULTS AND DISCUSSION

Electron Microscopy

After reduction in H_2 at 773 K, Pt is quite homogeneously dispersed, particle sizes being typically in the 10-20 Å range. Re-oxidation at 773 K (Fig. 1) does not affect significantly the distribution of the platinum phase, though the particles appear somewhat flatter, less contrasted and with a broader size distribution. The nature of such particles is presently studied under HRTEM conditions, to check whether the electron beam induces any reduction of the originally oxidized platinum. Noticeably, reduction of TiO_2 under the effect of the beam of the microscope was observed[4].

Hydrogen Adsorption and Desorption

The infrared spectra of the oxidized Pt/TiO_2 sample before (curve a) and after contact with 13.3 kNm^{-2} H_2 at 300 K for increasing time (curves b-g) are shown in Fig. 2. The main features in the background curve are: 1) a sharp peak at 1365 cm^{-1} due to surface sulfate groups[5] often introduced in titania catalysts

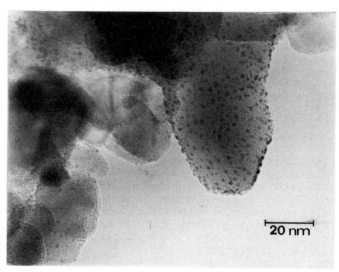

<u>Figure 1</u> Electron micrograph of Pt/TiO$_2$ heated in O$_2$ (13.3 kNm^{-2}) at 773 K.

during the preparation , 2) no surface OH bands are present (except an extremely weak band at 3720 cm^{-1}) and 3) severe scattering of the IR radiation with the transmittance falling to low values at highest frequencies, as constantly observed with dispersed systems.

As hydrogen is admitted (curve b) new intense bands appear monitoring the production of hydroxyls (OH stretching vibration at 3600-3720 cm^{-1}) and molecular water (H-O-H bending at 1615 cm^{-1}), both species being adsorbed on the surface of the TiO$_2$ support[5]. A complex and weak absorption (details in Fig. 3) due to stretching of hydride species on Pt particles[6] is observed at ≈ 2130 cm^{-1}. A very broad band due to the OH stretching modes of molecular water and hydrogen bonded OH groups is observed at 3000-3600 cm^{-1}. The bands described so far are shifted to low frequencies when D$_2$ is admitted instead of H$_2$. Finally, loss of transmittance is observed in the whole 4000-1050 cm^{-1} range in curve b. Upon increasing the time of contact with H$_2$ (Fig. 2, b-g), little or no increment of the discrete bands due to OH, H$_2$O and hydrides is observed, whereas the general decrease of transparency is progressively enhanced.

Hydroxyls and water are not removed from the surface when the pressure of hydrogen is reduced and upon outgassing at 300 K, whereas hydride bands decrease in the first stage of desorption and the overall transparency is progressively recovered almost up to the original values (Fig. 2,h). This effect is reversible, as re-admission of hydrogen restores curve g and the cycle may be repeated. None of the above effects are produced on pure TiO$_2$.

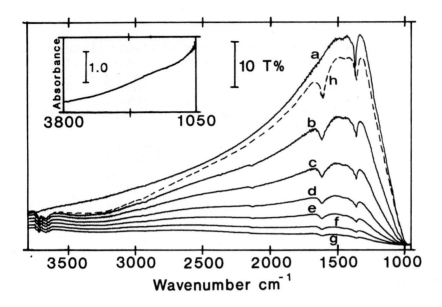

Figure 2 Effect of hydrogen on the oxidized Pt/TiO$_2$ sample. Infrared spectra: a) background, after heating in O$_2$ at 773 K and evacuation at the same temperature, b) immediately after the admission of 13.3 kNm^{-2} H$_2$ at 300 K; c-g) after 4, 9, 15, 30 and 60 min of contact with H$_2$ respectively; h) after outgassing 40 min at 300 K. Inset: absorbance curve obtained by subtracting spectrum h from spectrum g.

These data indicate that a) the oxidized platinum phase is reduced by hydrogen at room temperature, producing OH and water adsorbed irreversibly on the support; b) hydrogen is dissociated by the newly formed reduced metal phase and is partly stabilized on Pt^0 to give hydride species; c) some reversible reduction of the support occurs, causing the overall decrease of transmittance in the presence of removable hydrogen. As for this last effect, it is neither observed with not-reducible supports (e.g. SiO$_2$ and Al$_2$O$_3$)[2,6] nor with titania in the absence of Pt^0, suggesting that hydrogen, dissociated by the metal, spills over to the support, releasing electrons and finally populating the conduction band[2]. In fact, the dependence of the intensity decrease on the frequency (Fig. 2, inset) is typical of a light scattering phenomenon by free carriers in the conduction band[7]. Similar results[8] were obtained with Pt/TiO$_2$ reduced at 673 K. The reversible reduction of Ti^{4+} to Ti^{3+} ions by hydrogen at R.T. in Pt/TiO$_2$ was demonstrated by EPR measurements[2,9]. Such electronic effects associated with adsorbed species are relevant in gas sensor devices[3,10].

Hydride Species on Pt^0

Fig. 3 shows details of the spectrum, in expanded absorbance scale, of the hydride species at different hydrogen coverages. The various components in curve a are assignable to on top hydrides produced on differently coordinated Pt^0 sites[6]. A weak band at low frequencies is due to species more resistant to desorption at room temperature [6].

CO Adsorption Before and After Hydrogen Treatment

CO adsorption on the oxidized sample (Fig. 4,a) produces two main absorptions at 2130 cm^{-1} and at 2185 cm^{-1} with a much weaker component at ≈ 2210 cm^{-1}. The two latter bands are due to CO σ-bonded to acidic Ti^{4+} sites of the support[5] and that at 2130 cm^{-1} to CO on oxidized $Pt^{\delta+}$ sites[11,12]. By contrast, the spectrum of CO adsorbed on the sample reduced at 300 K by H_2 (curve b) is dominated by an intense and complex absorption at lower frequencies (2100-2000 cm^{-1}), in a range which is characteristic of the stretching mode of CO adsorbed on reduced Pt metal particles[11,12].

The contribution of oxidized platinum has practically disappeared. Also the bands of CO adsorbed on the support are strongly depressed due to the presence of hydroxyls and water produced during the reduction step.

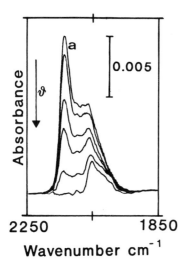

Figure 3 Absorbance spectra in the hydride stretching region at full (curve a, 13.3 kNm^{-2} H_2) and decreasing coverages (ϑ) by reducing pressure and outgassing at 300 K.

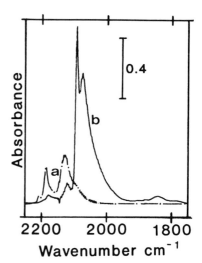

Figure 4 CO adsorption at 300 K. Infrared spectra (absorbance) immediately after the admission of CO (5.32 kNm^{-2}) on: a) the oxidized sample and b) the sample reduced by contact with hydrogen for 1 hour and outgassed at 300 K 40 min.

These results provide further evidence that: a) only oxidized platinum is exposed after the treatment in O_2; b) full reduction of the exposed platinum is attained by H_2 treatment at room temperature; c) water produced by reduction of the metal phase migrate onto the support. The presence of two components at 2092 and 2075 cm^{-1} monitors the surface heterogeneity of Pt particles[11,12] and, possibly, effects of coadsorbed hydrogen.

Acknowledgements. Assistance by Prof. M. Che in providing the samples and fruitful discussions are gratefully acknowledged, together with financial assistance by CEE and ASSTP (Associazione per lo Sviluppo Scientifico e Tecnologico del Piemonte).

REFERENCES

1. G.L. Haller and D.E. Resasco, "Advance in Catalysis", D.D. Eley, H. Pines and P.B. Weisz eds., Academic Press, San Diego, California, 1989, Vol. 36, Chapter 3, p.173.
2. W. Conner Jr., G.M. Pajonk and S.J. Teichner, "Advances in Catalysis", D.D. Eley, H.Pines and P.B. Weisz eds., Academic Press, San Diego, California, 1986, Vol. 34, Chapter 1, p. 1.
3. F. Boccuzzi, A. Chiorino, G. Ghiotti and E. Guglielminotti, Sensors and Actuators, 1989, 19, 119.
4. L. Wang, G.W. Qiao, H.Q. Ye, K.H. Kuo and Y.X. Chen, "Proc. 9th Int. Congr. Catal." M.J. Phillips and M. Ternan eds., Calgary 1988, Vol. 3, p. 1253.
5. C. Morterra, J. Chem. Soc., Faraday Trans. 1, 1988, 84, 1617.
6. T. Szilágyi, J. Catal., 1990, 121, 223.
7. V.F. Kiselev and O.V. Krylov, "Electronic Phenomena in Adsorption and Catalysis", Springer-Verlag, Berlin, 1987, Chapter 4, p. 67.
8. M. Komiyama and Y. Ogino, Colloids Surf., 1989, 37, 125.
9. J. C. Conesa and J. Soria, J. Phys. Chem., 1982, 86, 1392.
10. K.D. Schierbaum, U.K. Kirner, J.F. Geiger and W. Göpel, Sensors and Actuators B, 1991, 4, 87.
11. K. Tanaka and J.M. White, J. Catal.,1983, 79, 81.
12. L. Marchese, M.R. Boccuti, S. Coluccia, S. Lavagnino , A. Zecchina, L. Bonneviot and M. Che, "Structure and Reactivity of Surfaces", C. Morterra, A. Zecchina and G. Costa eds., Elsevier Science Publishers B. V., Amsterdam, 1989, p. 653.

X-Ray Absorption Spectroscopic Investigation of Microporous Crystalline Cobalt-substituted Aluminium Phosphates

J. Chen, G. Sankar, R. H. Jones, P. A. Wright, and J. M. Thomas

DAVY FARADAY RESEARCH LABORATORY, THE ROYAL INSTITUTION OF GREAT BRITAIN, 21 ALBEMARLE STREET, LONDON W1X 4BS, UK

1 INTRODUCTION

Among the crystalline microporous solids, transition metal substituted alumino phosphates (ALPOs) have been of recent interest, since the incorporation of heteroatoms other than Si, Al or P into the framework improves the catalytic performance of these materials[1,2]. Substitution of the framework aluminium ion in an ALPO by a divalent transition metal ion introduces a negative charge in the framework and as a result protons can be introduced into the structure to balance the charges. Cobalt has attracted considerable interest among the transition metal ions because it can exist in 2+ and 3+ oxidation states and can therefore act as a redox catalyst as well as Bronsted acid catalyst[1,3]. Due to the very low concentration and randomness of substitution of Co for Al it is not possible to determine unambiguously the position of the cobalt by conventional diffraction techniques, despite the high crystallinity of the samples. X-ray absorption spectroscopy (XAS), an atom-specific technique sensitive to the local environment of the metal ion and its oxidation state, is able to provide useful information on the coordination of the cobalt in these materials. Earlier Co K-edge XAS investigation[4] of CoAPO-20 suggested that the Co is distributed between both octahedral and tetrahedral coordination. In this paper we report Co K-edge XAS results of CoAPO-5, a large pore molecular sieve whose framework consists of uni-dimensional channels with 12 membered rings[5,6] (free diameter of 7.3Å). The investigations include the characterisation of the precursor, calcined and reduced materials.

2 EXPERIMENTAL

Details of the preparation of the Co-APO-5 have been reported in literature[7]. To remove the templates occluded in the framework structure the precursor was calcined in an O_2 atmosphere at 550°C for three hours. The calcined material was treated with methanol at 80°C, in order to investigate the nature of the Co ions in the material. The precursor, calcined and methanol-treated samples are designated as CoAPO-5p, CoAPO-5c and CoAPO-5r respectively.

Co K-edge XAS data were recorded at station 7.1 of the SERC Daresbury Laboratory. This station is provided with a Si[111] double crystal monochromator. CoO and Co_3O_4 were used as model compounds to extract the non-structural parameters. EXBROOK[8] (for pre and post edge background subtraction) and EXCURV90[8] (for curve fitting analysis) programs were used to analyse the EXAFS data. The XANES data were extracted after pre-edge subtraction and normalisation using the EXBROOK program. The error in determining the coordination numbers were ±10% and the interatomic distances were ±0.02Å.

3 RESULTS AND DISCUSSION

The Co K-edge XANES spectra of the precursor, calcined and methanol treated catalyst are shown in Fig.1. The XANES spectrum of the precursor shows a pre-edge feature

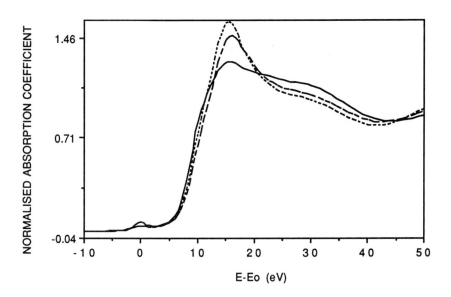

Fig.1 Co K-edge XANES spectra of precursor (————), calcined (- - - - - -) and the methanol treated (.....) CoAPO-5 sample.

(due to the 1s --> 3d transition) characteristic of Co in a tetrahedral environment. Similar observation was made in the Ni K-edge XANES of Ni^{2+}-substituted ALPO-5 in our earlier work[9]. The calcined and methanol-treated samples show a reduction in the intensity of the pre-edge feature indicating a change in Co environment. It is interesting to note that the edge position of the calcined catalyst shifted to higher energy by 4 eV compared to the precursor catalyst; the edge position of the methanol-treated sample shifts back to lower energy but remains higher than the precursor sample, indicating a partial reduction of Co^{3+} to Co^{2+}.

Analysis of the Co K-edge EXAFS data of the CoAPO-5p indicates that Co is surrounded by 4 oxygens at a distance of 1.95Å. The coordination number obtained confirm the XANES observation that Co is in a tetrahedral environment. The structural parameters obtained for the first shell are presented in Table 1. The EXAFS of the calcined sample is different from the precursor and analysis indicates the presence of oxygens at a short distance of 1.91Å and a longer distance of 2.09Å. The total coordination number of 5.5 oxygens at the two distances reflects the presence of tetrahedral and octahedral environments. A typical best fit obtained for the EXAFS data of the first shell is illustrated in Fig.2

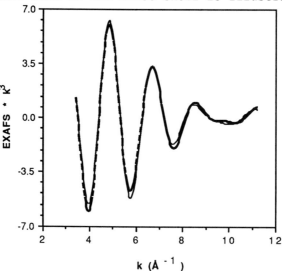

Fig.2. Typical best fit between the Fourier filtered (between 0.85 to 2.15Å) experimental data (solid line) of the calcined sample and the calculated EXAFS (dashed line) using structural data in Table 1.

and the structural data is listed in Table 1. The decrease in the pre-edge intensity in the XANES spectra could therefore be due to the presence of tetrahedral and octahedral environments as well as due to the higher mean Co-O distance[10,11].

Table.1 Structural parameters of the precursor, calcined and methanol treated CoAPO-5 catalyst

	Atom-pair	N	R(Å)	A(a)
Precursor	Co-O	4.1	1.95	0.009
Calcined	Co-O	2.6	1.93	0.009
	Co-O	2.9	2.09	0.010
Methanol treated	Co-O	2.4	1.95	0.008
	Co-O	2.8	2.10	0.008

(a) $A = 2 * \sigma^2$

The analysis of the EXAFS data of the methanol treated sample shows an increase in the Co-O distances whilst they retain the distorted environment. The increase in distance is consistent with the partial reduction of Co^{3+} to Co^{2+} inferred from the XANES spectra. The large value of the coordination number (5.2) and the distorted Co environment again explain the lower pre-edge intensity compared to that of the precursor sample.

Further detailed analysis of the data as well as characterization of these samples by other spectroscopic techniques are in progress.

4 ACKNOWLEDGEMENTS

We thank the Unilever PLC, SERC and The Royal Society for the support of this project.

5. REFERENCES

1. S.T. Wilson and E.M. Flanigen, Am. Chem. Soc. Symp. Ser. 1989, 398, 329
2. A. Meagher, V. Nair and R. Szostak, Zeolites, 1988, 8, 3
3. C. Montes, M.E. Davis, B. Murray and M.J. Narayana, J. Phys. Chem., 1990, 94, 6425
4. P. Behrens, J. Felsche and W. Niemann, Catalysis Today, 1991, 8, 479
5. J.M. Bennett, J.P. Cohen, E.M. Flanigen, J.J. Pluth, J.V. Smith, Am. Chem. Soc. Symp. Ser., 1983, 218, 109

6. S. Qiu, W. Pang, H. Kessler, J.L. Guth, Zeolites, 1989, 9, 440.
7. S.T. Wilson and E.M. Flanigen, US Patent, 1986, 4, 567,029.
8. Suite of programs is available at SERC Daresbury Laboratory.
9. Y. Xu, J.W. Couves,, R.H. Jones,C.R.A. Catlow, G.N. Greaves, J. Chen and J.M. Thomas, J. Phys. Chem. Solids, 1991, 52, 1229.
10 J. Wong, F.W. Lytle, R.P. Messmer and D.H. Maylotte, Phys. Rev. B, 1984, 30, 5596.
11. M. Nabavi, F. Taulelle, C. Sanchez and M. Verdaguer, J. Phys. Chem. Solids, 1990, 51, 1375.

Mechanism of Ethyne Hydrogenation in the Region of the Kinetic Discontinuity: Composition of C_4 Products

Richard B. Moyes[1], David W. Walker[1]*, Peter B. Wells[1], David A. Whan[1], and Elizabeth A. Irvine[2]

[1] SCHOOL OF CHEMISTRY, UNIVERSITY OF HULL, HULL HU6 7RX, UK
[2] ICI CHEMICALS AND POLYMERS LTD., PO BOX 90, WILTON, CLEVELAND TS6 8JE, UK

INTRODUCTION

The removal, by selective hydrogenation, of ethyne from the "ethylene" produced by steam cracking of naphtha is an important industrial process. Its aim is to hydrogenate ethyne to ethene, without hydrogenating either ethyne or ethene into ethane. The catalyst used commercially is palladium on a low surface area alumina support, and there have been several investigations into the fundamental aspects of the nature of the selectivity of this hydrogenation reaction [1 - 4]. We have previously reported results for the hydrogenation of ethyne, covering a variety of palladium catalysts, which show that there is a marked kinetic discontinuity on varying either the reaction temperature or the ethyne pressure [5]. The discontinuity is evident as a step in the Arrhenius plot at a temperature of about 370 K, between regions of high and low activation energy, and also as a step in rate, as a function of ethyne pressure, between regions of positive and slightly negative order in ethyne. The selectivity to ethene was also observed to fall abruptly as the temperature was increased through the transition point.

The hydrogenation of ethyne on palladium catalysts also yields products of oligomerisation. These are mainly dimers [6], although under commercial conditions higher degrees of polymerisation are readily observed as the so-called "green oil". The purpose of the present work is to explore the activity and selectivity of palladium, in terms of the yield and nature of the C_4-products from the dimerisation which accompanies the hydrogenation of ethyne, as the discontinuity is

*Present address: BP Research Laboratories, Sunbury-on-Thames.

traversed, in the expectation that it will shed light on the nature of the kinetic anomaly.

EXPERIMENTAL

The apparatus and methods used were identical to those already reported [5]. Briefly, reactions were carried out in a static reactor of volume 150 cm^3 and the extent of reaction was monitored by pressure fall. Hydrogen:ethyne ratios were in the range of 2:1 to 60:1. The nature of the products, after a chosen extent of reaction, was determined by gas chromatography.

As before [5], three catalysts were used. Catalyst A was 0.04 wt% palladium on a high area (gamma) alumina, catalyst B was 0.036 wt% palladium on a low area (alpha) alumina, and catalyst C was unsupported palladium powder produced by reduction of palladium chloride in hydrogen. The surface areas and other properties of the catalysts have been reported [5].

Catalyst samples were reduced in 200 Torr of hydrogen at 473 K for 2 h and then pumped for 0.5 h. Reactants were added to the vessel sequentially, the hydrogen being admitted first.

RESULTS

The variation with temperature of the yield of C_4 hydrocarbons for the three types of catalyst is presented in Figure 1a and the corresponding composition of the C_4 products is shown in Figure 1b. Attention is drawn to the following points: (a) the total yields of C_4 products decreased sharply at the transition temperature, (b) the production of butadiene fell markedly at the transition temperature, (c) the production of butane was minimal on the two supported catalysts, but on catalyst C it rose significantly at the transition temperature, and (d) the trans/cis ratio of but-2-enes did not vary greatly and was between 1.5 and 2.0 on all catalysts at all temperatures.

The effect of variation in ethyne pressure on the yield of C_4-hydrocarbons for the various catalysts studied is given in Figure 2a, and the accompanying product distributions are in Figure 2b. It is clearly seen that, as the ethyne pressure was raised through the discontinuity, the total C_4 yield became greater, the selectivity to butadiene increased substantially and, in the case of the palladium powder catalyst, the conversion to butane fell sharply. The trans/cis ratios of the but-2-enes were in general not particularly sensitive to the ethyne pressure, the value of ratio always lying between 1 and 2.

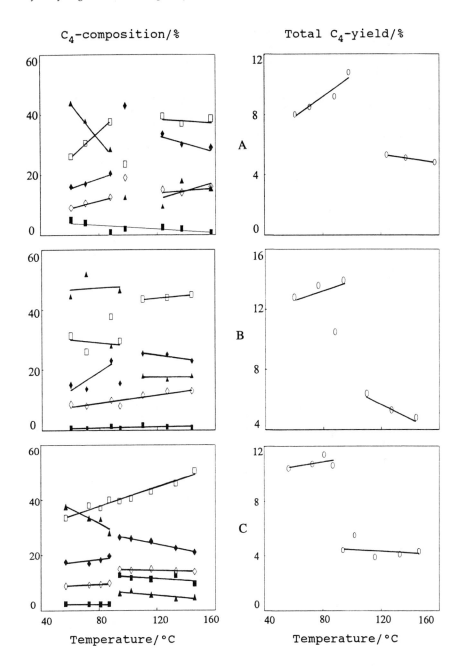

Figure 1. Variation with temperature of the compositions and yields of the C_4-hydrocarbons produced from reactions over catalysts A (top), B (middle) and C (bottom) for a hydrogen to ethyne ratio of 2:1. ■ , butane:
 □ , but-1-ene, ♦ , trans-but-2-ene; ◊ , cis-but-2-ene;
 ▲, butadiene.

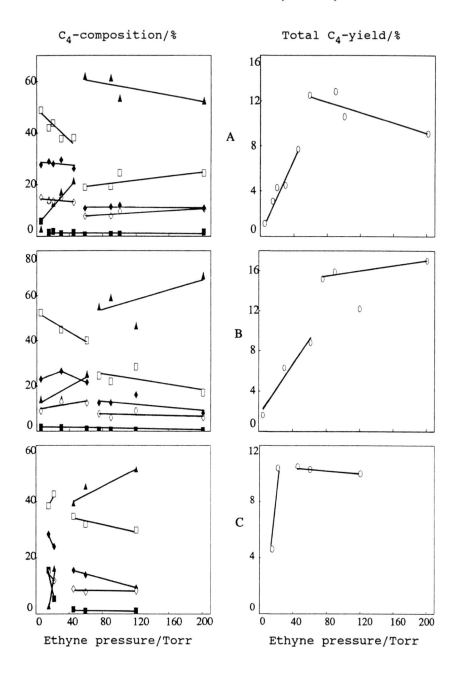

Figure 2. Variation with ethyne pressure of the compositions and yields of the C_4-hydrocarbons produced from reactions over catalysts A (top, 133°C), B (middle, 111°C) and C (bottom, 73°C). Symbols are as in Figure 1.

Experiments, carried out on catalyst A, to investigate the effect of hydrogen pressure (details not presented due to space limitation) showed that the overall yield of C_4 hydrocarbons fell from 16 to about 4% as the hydrogen pressure was raised through the discontinuous region, and this was accompanied by a fall in the relative amount of butadiene and an increase in the formation of but-1-ene and trans-but-2-ene.

DISCUSSION

The kinetic discontinuity in the rate of hydrogenation of ethyne on palladium catalysts has been interpreted in terms of a change in the packing of ethyne molecules on the surface of the catalyst [5]. This may be considered as a two dimensional phase change which, in common with other simple phase equilibria, is directly affected by both temperature and pressure. Our earlier results were consistent with the reasonable assumption that at higher temperatures, or lower ethyne pressures, the ethyne molecules are more widely spaced on the palladium surface than they are at lower temperatures or higher ethyne pressures.

The alterations herein reported at the discontinuity in the yield and nature of the C_4-products, which are produced in parallel with the hydrogenation of ethyne, may also be interpreted as consequences of the same change in the ethyne packing with temperature or pressure, and lend further support to a phase change in the adsorbed ethyne layer.

Consider first the sharp decrease in the total yield of C_4-products on raising the temperature or lowering the pressure through the discontinuity. The formation of C_4-hydrocarbons by dimerisation of ethyne will, almost certainly, be facilitated by circumstances in which ethyne molecules are closely packed on the surface of the catalyst, and our observations (see yields in Figures 1 and 2) are entirely consistent with this being so.

The extent of unsaturation in the C_4-products is also concordant with the concept of a surface phase change in ethyne. It is to be expected that, when ethyne packing density is high, hydrogen coverage will be low, and the formation of more highly unsaturated dimers favoured. Conversely, when the surface concentration of ethyne is low, and the hydrogen coverage is higher, the products of dimerisation will have a higher degree of saturation. Figures 1 and 2 clearly show that butadiene is indeed favoured, as compared with butenes, at lower temperatures and higher ethyne pressures. The amount of butane produced on both supported catalysts was too small for any conclusions to be drawn about its change on passing through the discontinuity. However, on the unsupported palladium powder the quantity of the fully saturated product increased, relative to unsaturated C_4-species, on passing to the regime where ethyne coverage was low and hydrogen coverage was high (Figures 1 and 2).

A study of the direct influence of hydrogen pressure on the character of the products of ethyne dimerisation was carried out only on catalyst A. The results, which are not reported, confirm that, as expected, the degree of unsaturation of the dimers decreased and the total yield of C_4 hydrocarbons fell as the hydrogen pressure was raised through the discontinuity.

CONCLUSION

The yields and compositions of the dimers co-produced during the hydrogenation of ethyne over palladium strongly reinforce the hypothesis of a two dimensional phase change in the packing of ethyne molecules on the metal surface as the temperature and pressure are altered.

ACKNOWLEDGEMENTS

We thank ICI Chemicals and Polymers Ltd and the Science and Engineering Research Council for the award of a CASE studentship to DWW.

REFERENCES

1. H. zur Strassen, Z. Phys. Chem., 1934, 81, 169.
2. G.H. Twigg and E.K. Rideal, Proc. Roy. Soc, 1939, A171, 55.
3. T. Tucholski and E.K. Rideal, J. Chem. Soc., 1935, 1701.
4. A. Farkas and L. Farkaas, J. Am. Chem. Soc., 1938, 60, 22.
5. R.B. Moyes, D.W. Walker, P.B. Wells, D.A. Whan and E.A. Irvine, Applied Catalysis, 1989, 55, L5.
6. G.C. Bond and P.B. Wells, J. Catalysis, 1966, 5, 65.

A Molecular Beam, RAIRS and NEXAFS Study of the Oxidative Dehydrogenation of Methanol on Cu(110)

R. Davis, S. M. Francis, C. Muryn, P. D. A. Pudney, A. W. Robinson, and M. Bowker
SURFACE SCIENCE RESEARCH CENTRE AND LEVERHULME CENTRE, DEPARTMENT OF CHEMISTRY, UNIVERSITY OF LIVERPOOL, LIVERPOOL L69 3BX, UK

ABSTRACT

Molecular beam reaction spectroscopy, reflection absorption infrared spectroscopy (RAIRS), and near edge x-ray absorption fine structure (NEXAFS) have been used to investigate methanol oxidation on Cu(110). The reaction is strongly dependent on the coverage of surface oxygen. For 1/4 monolayer the preadsorbed oxygen is all lost by reaction with methanol at 300K, leaving only the methoxy in a near upright configuration. When the saturation half monolayer of oxygen is predosed the adlayer is quite different after methanol dosing, giving a mixed layer of OH and methoxy species, both being observed in the RAIRS spectra.RAIRS and NEXAFS both indicate significant changes in the state of the methoxy on the surface either by orientational or charge transfer changes.

1 INTRODUCTION

The relationship between structure and reactivity of surfaces is a crucial one for determining how effectively a catalyst can convert reactants into products.This paper represents a small contribution to understanding these relationships and concerns one particular catalytic conversion, namely the production of formaldehyde by the oxidative dehydrogenation of methanol, in this case using a single crystal Cu(110) surface as the catalyst. The results of a combined kinetic (using molecular beam reaction spectroscopy) and structural (using RAIRS and NEXAFS) approach to understanding this reaction will be described.

2 EXPERIMENTAL

These experiments were conducted on three separate pieces of equipment. The kinetics and mechanism of the reaction were investigated with a thermal molecular beam system described in detail elsewhere[1,2]. It can be classified as a single collision reactor; the incoming molecules collide only once with the surface and are either reflected back into the gas phase (possibly after adsorbing and desorbing again quickly) or adsorb/react. Reactants and/or products coming off the surface are detected by a multimass quadrupole mass spectrometer. The new RAIRS system has a Mattson Galaxy 6021 FTIR spectrometer with a narrow band MCT detector, interfaced with a V.S.W. UHV chamber, the specially designed features of which will be described fully elsewhere[3]. The spectra presented here are the the result of 1000 scans from the clean background surface ratioed to 1000 scans after adsorption at $4cm^{-1}$ resolution. The NEXAFS experiments employed the high-energy spherical grating monochromator ($200<h\nu>1000$ eV) on beamline 1.1 at the SRS, Daresbury Laboratory [4], using an experimental chamber described elsewhere [5]. The O K edge NEXAFS spectra were recorded using Auger detection, monitoring the yield of O KVV electrons at a kinetic energy of 505 eV and scanning the photon energy range 520-560 eV. At the C-1s edge due to the 40% higher order light content and the 36% absorption feature, only 24% of the light is available for the NEXAFS experiment. As the signal is on such a high background this leads to problems in the normalisation procedure introducing artefacts into the normalisation process for the spectra, so our discussions will concentrate on the O 1s edge.

3 RESULTS AND DISCUSSION

The results of the molecular beam experiments have been described in detail elsewhere[2], but it is important to summarise these data here to put the structural work into context. Fig.1 shows a comparison between the results for methanol reaction with the Cu(110) surface predosed with 1/4 and 1/2 monolayer of oxygen atoms. The reaction kinetics are quite different in these two circumstances, though the overall mechanism is the same, that is:

$$2 CH_3OH + O_a \rightarrow 2 H_2CO + H_2 + H_2O$$

With 1/4 monolayer of oxygen there is a 'clean-off' reaction with immediate and rapid water production, leaving a clean surface at completion. This occurs due to the rapid decomposition of the main intermediate in this reaction, the methoxy species. In fact, if the adsorption is carried out at 300K the methoxy is quite stable and subsequent TPD shows decomposition occurring at around 370K[6,7]. This represents the stability of methoxy on clean Cu, since all the oxygen was lost as water into the gas phase. In contrast, the data in fig 1 show an induction time to formaldehyde evolution with 1/2 monolayer of preadsorbed oxygen atoms, and this was shown by reaction modelling to be due to an enhanced stability of methoxy in these circumstances. This is due to the fact that there is no clean-off of the preadsorbed oxygen; the induction period is due to the following reaction going to complete conversion before water formation:

$$CH_3OH + O_a \rightarrow CH_3O_a + OH_a$$

In this coadsorbed state, adjacent vacant Cu sites are blocked, thus poisoning the decomposition reaction and conferring enhanced stability on the methoxy. Product formation only occurs (and accelerates) when vacant sites in the OH adlayer begin to occur (perhaps at more reactive step or defect sites), thus inducing decomposition, H_a formation and water production (hence freeing up more sites). The model for the stabilised adlayer in this situation is as shown in fig.2, with an ordered mix of methoxy and hydroxy species (LEED shows a p(2x1) pattern in this case).

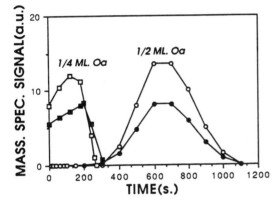

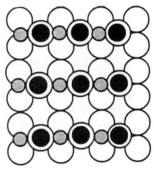

Figure.1. Formaldehyde (open points) and water (filled points) evolution during beaming of methanol onto the Cu(110) surface predosed with 1/4 ML (squares) at 333K and 1/2 ML (circles) at 353K of oxygen atoms. Hydrogen evolution is not shown, but is coincident in time profile with other products.

Figure.2. Model of Cu(110) surface after formation of the mixed methoxy(solid filled circles), hydroxy(dotted filled circles) adlayer after preadsorption of 1/2 ML of oxygen atoms. The latter are converted to the hydroxy group at the original site. The methoxy species is shown adsorbed on top of the copper atoms in the missing row structure which is on top of the (1x1) Cu(110) lattice.

The RAIRS results were obtained by preadsorbing 1/4 or 1/2 monolayers of oxygen and then subsequently adsorbing methanol to produce surface species and are shown in Fig.3. The spectra for the 1/4 ML are very similar to those reported for the methoxy intermediate on Cu(111) [8] and Cu(100) [9] by RAIRS, that is three bands are observed in the CH stretching region and one other band is also observed, at 1024 cm^{-1}, assigned to the CO stretch, but with the CH bending vibrations absent. The EELS spectra have been reported on Cu(110)[10] and are consistent with these observations but with the methyl deformations and rock also observed. The two RAIRS studies had different conclusions for the orientation of the methoxy species on the surface, Ryberg[9] concluded that that the methyl group was at an angle to the surface on Cu(100) due to the observation of the asymmetric methyl stretches, in contrast with Chesters et al [8] who propose that the species is perpendicular to the surface on Cu(111) arguing that one of the assymetric modes is not allowed under the surface selection rule and also the frequencies of the bands assigned to these modes are unrealistic. Spectra of the deuterated analogue showed the two modes assigned to asymmetric stretches by Ryberg are most likely to be overtones, being at 2922cm^{-1} and 2888cm^{-1} on Cu(110) with 1/4ML preadsorbed oxygen.The other band in the CH stretching region (at 2815 cm^{-1} in our case) is the symmetric methyl stretch. Thus our results for the 1/4 monolayer on Cu(110) support the methoxy being upright or close to upright.

The spectra obtained by preadsorbing 1/2 ml of oxygen are shown in fig 3 b&c, and some significant changes from the 1/4 ml spectra can be seen. A new broad band is seen at approximately 3150 cm^{-1} and weak new features in the CH stretching region. There are also some changes to the bands also present in the 1/4ml case ,notably the shift of the symmetric methyl stretch from 2815 cm^{-1} to 2805cm^{-1} and the carbon oxygen stretch from 1024cm^{-1} to 1035cm^{-1} with all the bands now observed generally broader than in the 1/4 ml case. The most significant of these observations is that of the new band at the higher frequency of 3150cm^{-1}, this is lower than the ice band which grows on the liquid nitrogen cooled MCT detector, which is shown at a similar intensity from two clean surface background spectra ratioed together in Fig.3.d. The 3150cm^{-1} band is thus assigned to an OH stretch on the surface produced from the hydrogen from the methanol reacting with the preadsorbed oxygen to produce OH on the surface coadsorbed with the methoxy species, as suggested from the results from our molecular beam work discussed above. The changes to the spectrum of the methoxy species, broadening of the bands and the shift of the $\upsilon_s CH_3$ and υCO modes, also indicate the possibility of some interaction between the species present on the surface which could give rise to the enhanced stability of the the methoxy species observed in the molecular beam results. Alternatively there may be some change in orientation of the methoxy species due to crowding on the surface. Further work is presently being undertaken to confirm and extend these early results will to be presented in full at a later date.

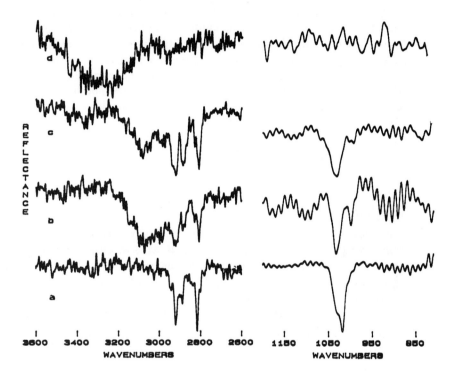

Figure 3. Fourier Transform RAIRS spectra of methoxy species on Cu(110) a). 1/4 monolayer of preadsorbed oxygen and 4L of methanol heated to 283K. b). 1/2 monolayer of preadsorbed oxygen and 4L of methanol heated to 283K c). repeat experiment as in b). d). clean spectra ratioed together.

The NEXAFS spectra were normalised by dividing each adsorbate covered spectrum by a clean surface spectrum and standardizing the edge jump to 1. Figure 4 shows a series of normalised NEXAFS scans at the O-edge for the methanol layer adsorbed onto a 1/4 monolayer dose of oxygen. The angle of incidence of the photons was varied relative to the [001] (fig 4a) and the [110] azimuth (fig 4b). Both show the same features, a small resonance just above the edge (A), present at normal but not grazing incidence, and a much broader resonance (B) further from the edge which increases in intensity from normal to grazing incidence. It is generally accepted that for CH_3O on Cu(110) the molecular axis lies closer to the surface normal than to the surface [11,12]. The CH_3O molecule has C_{3v} effective symmetry [13,14], the unfilled molecular orbitals are the $6a_1$, $7a_1$ and 3e, the $6a_1$ orbital having the lowest energy (fig.5). The polarisation dependence of the a_1 orbitals would be to increase in intensity from normal to grazing incidence. For methoxy on Cu(100) and (111) previous workers [13,14] have assigned the $6a_1$ orbital to be a component of the broader resonance (B). The resonance just above the edge (A) has the polarisation dependence expected for an e type orbital but since the 3e orbital is expected to have much higher energy then the $6a_1$, a possible assignment is that it is due to the 2e orbital. For $CH_3O^.$ the 2e is both the HOMO and LUMO; however for CH_3O^- the 2e is the HOMO and the $6a_1$ is the LUMO. The 2e orbitals could hybridise with the Cu d-band making it possible to have 2e-type states above the valence band, so if charge transfer to the $CH_3O^.$ does not form a fully negative ion, the 2e orbital is partially filled and so visible in NEXAFS as a small resonance just above the edge with maximum intensity at normal incidence. The energy separation of the 2e and $6a_1$ levels for $CH_3O^.$ is 4.19eV, whereas for the CH_3O^- it is 11.53eV [15]. The energy separation of the 2 resonances A and B is 4 eV for the [001] azimuth and 5 eV for the [110] azimuth. Therefore the assumption that A is the 2e orbital seems reasonable. The 2e orbitals are essentially the lone pair of oxygen p orbitals in the plane of the surface, the p-lobes will either run along the rows, or towards the rows, so would expect differences in resonance A for the two azimuths, which is what is observed. This preliminary assessment of the data indicates that for a 1/4 monolayer predose of oxygen the CH_3O axis lies close to the surface normal. The NEXAFS spectra for 1/2 monolayer oxygen prior to CH_3OH dosing are shown in figure 6. The spectra from both azimuths were essentially identical, only one broad resonance above the edge which shows very little angular dependence. The resonance will

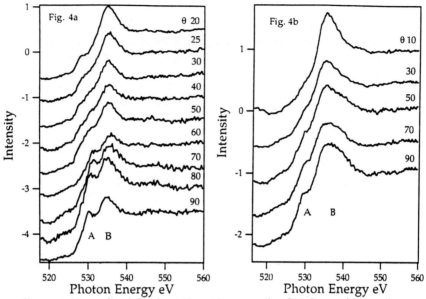

Figure 4. Normalised O-edge NEXAFS spectra for CH_3O on 1/4 monolayer oxygen predosed Cu(110), a. [001] azimuth, b. [110] azimuth

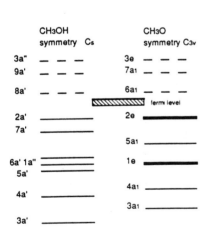

Figure 5. Schematic of the molecular orbital levels of methanol and methoxy.

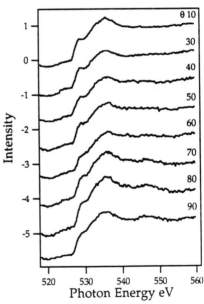

Figure 6. Normalised O-edge NEXAFS spectra for CH_3O on 1/2 monolayer oxygen predosed Cu(110), [001] azimuth.

have contributions from OH as well as CH$_3$O unfilled orbitals and possibly features due to substrate scattering [4]. Therefore to obtain the orientation information C-edge data is required. It is interesting to note that the small feature just above the edge assigned to the 2e orbital on the O-edge for the 1/4 monolayer oxygen predose is absent for the 1/2 monolayer oxygen predose, indicating the presence of OH has altered the charge transfer to the CH$_3$O, possibly causing the 2e orbitals to become filled and hence no longer probed by NEXAFS.

In summary both spectroscopic and kinetic techniques indicate a strong oxygen coverage dependence of the adlayer formed from methanol adsorption. In particular a stable coadsorbed layer of CH$_3$O and OH is formed when a saturated layer of oxygen atoms is initally present.

REFERENCES

1. M.Bowker, P.D.A.Pudney, C.J.Barnes. J.Vac.Sci.Technol. A, 1990,8,816
2. C.Barnes, P.D.A.Pudney, Q.Guo, M.Bowker. J.Chem.Soc. Faraday. Trans. 1990,86,2693
3. P.D.A.Pudney, M.Bowker. To Be Published.
4. M. Surman, I. Cragg-Hine, J. Singh, B. Bowler, A. H. Padmore, D. Norman, R. Davis, K. G. Purcell, G. Thornton, A. L. Johnson, A. Atrei, W. K. Walter, D. King. Rev. Sci. Instrum. In press.
5 R. Davis, R. Lindsay, K. G. Purcell, G. Thornton, A. R. Robinson, T. P. Morrison, M. Bowker, P. D. A. Pudney, M. Surman. J. Phys. Condens. Matter ,1991,3 ,297.
6. D.Ying, R.J.Madix. J.Catal.,1980,61,48.
7. M.Bowker,R.J.Madix. Surf.Sci.1981,102,542.
8. M.A.Chesters,E.M.McCash. Spectrochimica Acta 1987,43A,1625.
9 R.Ryberg, Phys.Rev.B., 1985,31,2545.
10 B.Sexton,A.E.Hughes,N.R.Avery. Surf.Sci. 1985,155,366.
11 K. C. Prince, E. Holub-Krappe, K. Horn, D. P. Woodruff. Phys rev.B 1985 32, 4249.
12 M. Bader, A. Puschmann, J. Hasse. Phys Rev B 1986,33, 7336.
13 Th. Lindner, J. Somers, A. M. Bradshaw, A. L. D. Kilcoyne D. P. Wooddruff. Surf Sci.1988,203, 333.
14 D. E. Ricken, J. Somers, A. Robinson A. M. Bradshaw. Faraday Discuss. Chem. Soc.1990,89,291.
15 J. A. Rodriguez, C. T. Campbell. Surf Sci. 1988,194, 475.

Ethanol Oxidation on Rhodium(110)

Y. Li and M. Bowker
SURFACE SCIENCE RESEARCH CENTRE AND LEVERHULME CENTRE,
DEPARTMENT OF CHEMISTRY, UNIVERSITY OF LIVERPOOL,
LIVERPOOL L69 3BX, UK

ABSTRACT The oxidation of ethanol on the Rh(110) surface has been followed using molecular beam reaction spectroscopy and temperature programmed desorption. The dehydrogenation pathway seen on the clean surface is significantly modified by the stabilization of surface intermediates such as methyl groups (which are hydrogenated efficiently to CH_4) and more importantly the acetate intermediate. This is stabilized by ca 150K in the presence of coadsorbates and it decomposes in an autocatalytic manner which can be classified as a 'surface explosion'.

1 INTRODUCTION

It is well known that rhodium is the most versatile transition metal to catalyse the hydrogenation of CO, with a diverse product spread of hydrocarbons and alcohols depending upon the exact state of the surface[1]. Of significant industrial interest is the synthesis of ethanol on rhodium catalysts, but the mechanism of this has long been a controversial topic. Bowker has reviewed this subject[2], and a mechanism with acetate as an important intermediate has been given. In one of our previous papers[3] acetate formation was observed during ethanol oxidation on the Rh(110) surface, and the mechanism of the decomposition of this species was shown to be an unusual one, namely it exhibited a 'surface explosion'. These kinds of reactions were first observed by Madix's group[4,5], by heating a formic acid or acetic acid adsorption phase on Ni(110). In this work the oxidation of ethanol on Rh(110) is explored in detail, with the aim of fully describing the effect of oxygen on ethanol dehydrogenation on rhodium.

2 EXPERIMENTAL

The techniques we have used are molecular beam reaction spectroscopy and temperature programmed desorption (TPD). The equipment and methodology were described in detail elsewhere[6,7]. The molecular beam system can be classified as a **single collision reactor**, since a molecule in the beam hits the surface only once and is either reflected or desorbed again, which will be detected in the mass spectrometer(MS), or it sticks on the surface. The latter can lead to a reaction and the products evolved can be detected by the MS in-situ. By monitoring the MS intensities of the relevant species, the adsorption or consumption of the adsorbent and the formation of the reaction products can be closely followed. Subsequent TPD experiment can provide more information on the reactions on the surface, and this technique provided the evidence for the formation of acetate during ethanol oxidation[3].

The sample is a rhodium single crystal disc oriented in the (110) direction, which was cleaned by cycles of argon ion bombardment - annealing - oxidation and CO treatment until a sharp and clean (1X1) LEED pattern was achieved and an Auger spectroscopic measurement revealed only rhodium signals. Every time before beaming the cleanliness of the sample was checked by oxygen or CO adsorption, and if there was any doubt LEED and Auger would be used. Oxygen was first beamed onto the surface, then it was removed from the dosing volume, replaced by ethanol and the adsorption restarted by unblocking the beam and allowing it to impinge onto the surface. Temperature programmed desorption also utilised the mass spectrometer for product detection and a heating rate of 1.5 K s^{-1} was used. Both oxygen and ethanol were beamed with the sample at ca 315K, unless otherwise stated.

3 RESULTS AND DISCUSSION

Within two minutes of oxygen beaming the oxygen MS signal is constant, indicating that the rhodium surface is saturated with oxygen. According to the previous work of this group the oxygen coverage at this stage is about 0.7 monolayer of atoms[8]. Onto this kind of surface ethanol adsorbed readily, and saturation was reached within two minutes[3]. Figure 1A shows the molecular beam reaction spectra of the consumption of ethanol and the evolution of the reaction on an oxygen saturated surface, with methane, hydrogen and water as the products. Afterwards a TPD run produces the unusual explosive decomposition of acetate with very narrow peaks of hydrogen and CO_2 at 390K (figure 2), which was discussed in detail in a previous publication[3]. In order to clarify the kinetics and mechanism of the reaction, we will present here three sets of experiments to elucidate the effects of the variation of oxygen pre-coverage, the variation of ethanol dosing, and the variation of ethanol dosing temperature.

To consider the oxygen coverage effect excess ethanol was beamed to surfaces predosed with different

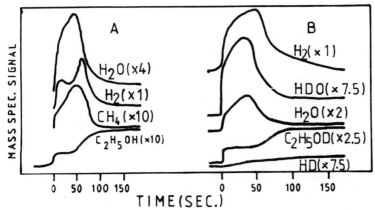

Figure 1 Ethanol adsorption onto oxygen saturated Rh(110).
A. Normal ethanol. B. Deuterated ethanol.

amounts of oxygen. With small oxygen coverages the main product observed during adsorption was hydrogen, with a small amount of water released upon ethanol introduction. For increasing oxygen doses, the water peak increased (always at the admission of the beam), the hydrogen maximum came later with diminished peak. Only when the sample was predosed with oxygen for 70 seconds or more, which corresponds to ca 90% of saturation or > 0.6 monolayer, did the subsequent TPD experiment produce the explosive decomposition of the acetate (figure 2). The reaction proceeds as follows. If there is only a limited amount of oxygen available, beamed ethanol can react with surface oxygen to form hydroxyl group and ethoxy, but the ethoxy is unstable and will decompose fast to form carbon, surface CO, and hydrogen. Part of the hydrogen also reacts fast with surface hydroxyl to form water.

$$C_2H_5OH + O(a) \rightarrow CH_3CH_2O(a) + OH(a) \quad (1)$$
$$CH_3CH_2O(a) \rightarrow C(a) + 5H(a) + CO(a) \quad (2)$$
$$2H(a) \rightarrow H_2(g) \quad (3)$$
$$H(a) + OH(a) \rightarrow H_2O(g) \quad (4)$$

At the same time hydrogen can also scavenge oxygen:

$$H(a) + O(a) \rightarrow OH(a) \quad (5)$$
$$H(a) + OH(a) \rightarrow H_2O(g) \quad (6)$$

However if more oxygen is available, some other reaction may take place in addition to those mentioned above. Some ethoxy may be turned to acetate by oxygen attack at the α carbon, which is not so fast as step 4 at the temperature used, and so requires higher oxygen coverage, such that the ratio between H(a) and O(a) is beyond stoichiometric water formation.

$$CH_3CH_2O(a) + O(a) \rightarrow CH_3COO(a) + 2H(a) \quad (7)$$

It is clear from fig.1 that some ethoxy may undergo

another kind of reaction to produce methane.

$$CH_3CH_2O(a) \rightarrow CH_3(a) + 2H(a) + CO(a) \quad (8)$$
$$CH_3(a) + H(a) \rightarrow CH_4(g) \quad (9)$$

It appears that the methyl groups are stabilized in the presence of concentrated surface oxygen, perhaps because of a site-blocking effect on the methyl decomposition pathway, and because of their increased lifetime the methyl can recombine with a hydrogen atom to form methane and leave the surface before further decomposition.

To support this mechanism, some experiments were carried out using deuterated ethanol, CH_3CH_2OD. The results are shown in figure 1B. During ethanol beaming onto the oxygen saturated surface both H_2O and HDO evolve with a ratio of about 2:1. The equivalent steps for these are (4), (5) and (6). No trace of D_2O is found. As with normal ethanol, hydrogen is produced too, while HD and D_2 are totally missing. This is consistent with our mechanism very well, as hydrogen molecules are formed by atoms other than the hydroxyl ones.

It must be mentioned that neither CO nor CO_2 are evolved during ethanol adsorption, which is a little surprising since step (2) gives rise to CO(a), and this will react readily with surface oxygen at the temperature used[8]. Our explanation for this is that CO is localized to the reaction site of step (2), and because of the fast nature of steps (1) and (5) the oxygen nearest to CO is used up well before CO oxidation can happen. It may also be that adjacent oxygen is present as OH and this has a much faster reaction rate with H(a), compared with CO(a). Therefore CO remains on the surface until the sample is heated up to ca 450K.

If the oxygen predosing period was fixed to 5 minutes (saturated) and ethanol beaming time varied, all the above discussion of ethanol adsorption onto a highly concentrated oxygen precovered surface holds, but the process of acetate decomposition varies. With 15 and 30 seconds of ethanol dosing the explosion products were water and CO_2 exclusively (fig.2A) while with 50 and 75 second dosings hydrogen replaced water as the major product (fig. 2B), and the water peak shrank to a very low level. Obviously the condition favours the oxygen attack to form water when a smaller amount of ethanol is admitted, since the O/ethanol ratio is higher than with excess ethanol dosing. Towards the end of dosing there is still oxygen available on the surface. The acetate is stabilized by the presence of oxygen and decomposes at around 390K, and the released hydrogen atoms react with the surface oxygen to form water. It is reasonable that when more ethanol is dosed, less oxygen survives to react with hydrogen, so that hydrogen replaces water as the dominating decomposition species.

Ethanol Oxidation on Rhodium(110)

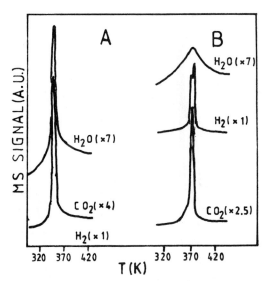

Figure 2 TPD after ethanol beaming onto oxygen saturated Rh(110). A. 30 seconds. B. 75 seconds.

The shape of the acetate decomposion profile was described in detail previously[3]. It was rationalized that an explosion takes place if the reaction requires a specific site and that more such sites are created during the reaction. Based on this the reaction rate was given as

$$R = 2kD/N_s^{1/2} \, [\theta_N \cdot \pi \cdot a(1-\theta)]^{1/2}, \tag{10}$$

where k is the rate constant, D is the density of reactive sites at the circumference of a reactive patch, N_s is the total number of sites involved, θ_N represents the coverage of nucleation sites, a is the beam area, and θ is the coverage of acetate. As an inference of this, if the ethanol dosing temperature is fixed at near but below the temperature of the explosion maximum, the acetate decomposition will begin and accelerate due to its autocatalytic nature. Keeping this in mind we carried out a set of experiments with ethanol beaming onto an oxygen-saturated surface at various temperatures, and the results are shown in figure 3. 345K is too low to initiate the explosion (fig.3A). At 361K the explosion peaks around 25 seconds after the dosing starts (fig.3B). This shortens at higher temperatures: 15 seconds at 374K (fig.3C), 9 seconds at 395K (fig.3D), and 5 seconds at 423K (fig.3E). Further temperature increase to 573K (fig.3F) results in a change of kinetics, and an induction period for the desorption of water, CO_2, and hydrogen, which is much like the oxidation of methanol on Rh(110)[9].

Interestingly the arrangement of surface rhodium atoms plays an important role on the ethanol oxidation. Barteau's group has studied the reaction on a Rh(111) surface[10], and they observed the formation of acetate, but

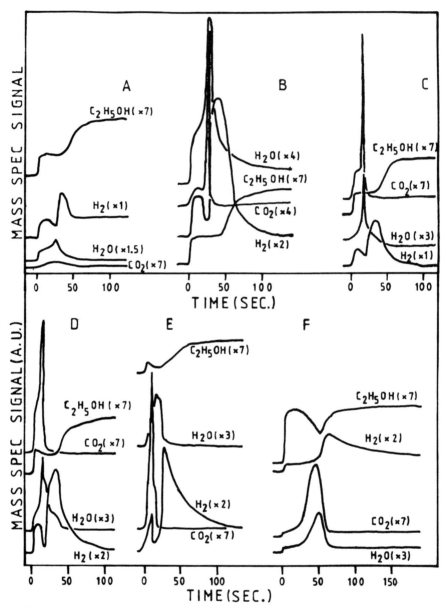

Figure 3 Ethanol beaming onto oxygen saturated Rh(110) at various temperatures. A. 345K. B. 361K. C. 374K. D. 395K. E. 423K. F. 573K.

no 'surface explosion' has been found. Our preliminary results on Rh(111) also indicate that a more close-packed surface is less active for the oxidation of ethanol, and for the 'surface explosion'.

One more thing worth mentioning is the adsorption and

decomposition of acetic acid on the same Rh(110) sample. The TPD experiment after acetic acid dosing revealed similar explosive desorptions of CO_2 and hydrogen at the same temperature range. Pre-covering of carbon, nitrogen and oxygen could greatly modify the stability of the surface acetate, to increase or decrease the decomposition temperature. As a result the explosion takes place at various temperatures. This will be discussed in another paper[11].

REFERENCES

1. For a review see: E.Poels and V.Ponec, Catalysis, 1983, 6, 196.
2. M. Bowker, Catalysis Today, in press.
3. M. Bowker and Y. Li, Catalysis Letters, 1991, 10, 249.
4. R. J. Madix, J. L. Falconer and A. M. Suszko, Surface Science, 1976, 54, 6.
5. J. L. Falconer and R. J. Madix, Surface Science, 1974, 46, 473.
6. M. Bowker, P. Pudney and C. Barnes, J. Vac. Sci. Tech., 1990, A8, 816.
7. C. Barnes, P. Pudney, Q.Guo and M.Bowker, J. Chem. Soc. Farad. Trans., 1990, 86, 2693.
8. Q. Guo, R. W. Joyner and M. Bowker, J. Phys.: Condens. Matter 1991, 3, S55.
9. M.Bowker, Q.Guo and R.Joyner, Surf. Sci. 1991, 253, 33.
10. C. J. Houtman, Ph.D Dissertation, University of Delaware, 1990
11. M. Bowker and Y. Li, to be published.

Atom Probe Field Ion Microscope Study of Surface Changes in Platinum/Rhodium Catalyst Alloy upon Oxidation and Reduction Treatments

Stephen Poulston[1] and George D. W. Smith
DEPARTMENT OF MATERIALS, UNIVERSITY OF OXFORD,
PARKS ROAD, OXFORD OX1 3PH, UK

1 INTRODUCTION

The technique of Atom Probe Field Ion Microscopy (APFIM) has a number of features which commend its use in the analysis of catalyst surfaces. These include: the close resemblance of the specimen end form to a single supported catalyst particle in both size and morphology; its ability to produce true atomic scale elemental depth composition profiles, removing material from the surface atomic layer by atomic layer, in a regular and controlled manner; its ability to produce statistically highly accurate elemental analyses even from the small number of ions collected by the removal of monolayers of material. A full explanation of the technique is available elsewhere.[1,2]

APFIM has been successfully applied to the investigation of surface segregation problems in a number of systems. A review of this work has been produced.[3] One system which has excited considerable interest is the Pt/Rh alloy which is used extensively as a pollution control catalyst most notably in automobile exhaust emission control[4] and as an industrial catalyst in processes such as ammonia oxidation. From a technical point of view also Pt/Rh alloys are well suited to study by FIM. This is a result of the similar evaporation fields required for Pt and Rh.[1] Artifacts in the compositional analysis generated by preferential evaporation of one of the metals in the alloy are therefore avoided.

Vacuum annealing of Pt/Rh alloys is found to produce a Pt enriched surface,[5] followed by a Pt depleted layer.[6] However, oxidation of this material produces a Rh enriched surface[7,8] and this has been reported to be maintained under reduction conditions sufficient to regenerate the metal surface.[9] Though this latter investigation produced information on the Rh enriched layer it did not include information on the subsurface region.

The present study provides, for the first time, true atomic scale elemental depth profiling of a Pt/Rh alloy after exposure to a range of oxidation and reduction treatments. The depth composition

---

1. Present address: Department of Chemistry, University of Cambridge, Lensfield Road, Cambridge. CB2 1EW.

profiles presented here were obtained using a Position Sensitive Atom Probe (PoSAP), a unique variant on the conventional AP.[10] The PoSAP employs a far larger probe aperture which enables a greater number of atoms to be collected from each atomic layer of material removed from the sample surface. This produces compositional information with a far greater statistical accuracy than is possible with the conventional AP. Also presented are some FIM and Transmission Electron Microscope (TEM) micrographs of typical specimens showing the effect of the oxidation and reduction treatments on the apex region.

2 EXPERIMENTAL

The material used was a Pt/17 at.% Rh alloy wire of diameter 0.2mm supplied by Johnson Matthey p.l.c. This composition is the same as the typical bulk composition of the noble metal content in catalytic converters.

For a specimen to be examined in the AP it must have a needle-like end form with a radius of curvature of 1000Å or less. This is achieved by an electropolishing process using a piece of wire approximately 1 cm long with an electrolyte consisting of molten $NaCl/NaNO_3$ in the ratio 1:4 by volume. This procedure has been described elsewhere.[11] The specimen was then imaged and field evaporated in a FIM at a temperature of 80 K in 5×10^{-5} Torr of neon. The field evaporation was continued until a full even image was obtained (Figure 1a).

The specimen was then removed from the FIM for oxidation. This was carried out in air at 1 atm. pressure using a tube furnace. Specimens were oxidised at 600°C and 800°C for a period of 1 hour before being removed from the furnace to air cool. Some specimens were retained for examination whilst others were reduced, again in a tube furnace, using a stream of H_2 at 1 atm. pressure.

The apex region of an APFIM specimen is sufficiently thin to allow examination in a high voltage (greater than around 100KV) TEM without any further treatment. Typical examples of specimens were inspected in a Phillips CM 20 TEM. Figure 1b is a transmission electron micrograph of an untreated specimen after it has been field evaporated. It shows an even surface finish with a gradual taper to a sharp end point.

The heat-treated specimens were then inserted into the AP and imaged as described above. The voltage applied to the specimen was raised to around 4KV, by which point only a few bright spots were visible in the image. This was sufficient to align the specimen in relation to the detector. The imaging gas was then removed and atom probe analysis begun with a pulse voltage set at a constant fraction of 15% of the applied standing voltage.

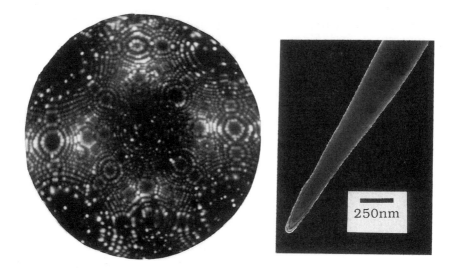

Figure 1:a (left) Field ion micrograph and b (right) Transmission electron micrograph of a Pt/Rh alloy tip after field evaporation.

3 RESULTS

a) TEM Examination. The untreated specimen in figure 1b shows an even, rounded end form as a result of the field evaporation process described above. The end form has a radius of curvature of approximately 250 Å. It can be seen therefore that the end form compares well both in size and topography to a single supported catalyst particle. Figure 2a shows a specimen which has been oxidised at 800°C for 1 hour. The smooth end form is no longer observed with the specimen being highly uneven and showing signs of faceting. Although the entire surface appears covered in oxide several patches can be seen where the oxide seems to be especially thick. Figure 2 b shows a specimen which has been oxidised at 800°C for 1 hour and then reduced in H_2 at 400°C for 30 minutes. The surface finish appears smoother than in 2a with a more regular end form.

b) FIM Examination. Figure 3a shows a FIM micrograph of a specimen which has been oxidised at 800°C for 1 hour. The micrograph was taken at 8.25KV and reveals two distinct areas on the specimen surface. One region, characterised by an irregular array of bright spots, is produced by the oxide coating on the specimen surface. The other region shows a regular pattern consisting of concentric circles of finer spots and this represents the clean metal emerging through the oxide coating. Figure 3b shows a FIM micrograph of a specimen oxidised at 600°C for 1 hour. This is taken at a similar voltage of 8.3KV. In this case, however, the distinct regions of oxide and metal visible in 3a are not apparent with the metal emerging through the oxide coating in a far more uniform manner over the entire surface. FIM observations of oxidised and reduced specimens showed no signs of surface oxide demonstrating that the reduction process had been successful in regenerating the metal from the oxide coating.

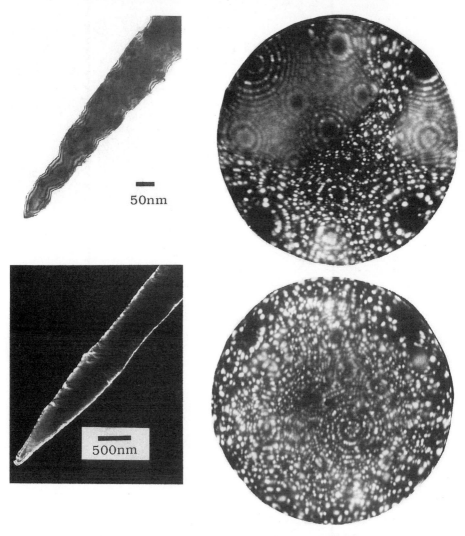

Figure 2: Transmission electron micrograph of a FIM tip a (top) oxidised at 800°C in air for 1 hour at 1 atm. b (bottom) oxidised at 800°C for 1 hour in air then reduced at 400°C in H_2.

Figure 3: Field Ion Micrograph of a FIM tip oxidised at a (top) 800°C b (bottom) 600°C in air at 1 atm.

c) AP Microanalysis. An elemental depth composition profile for a specimen oxidised at 800°C for 1 hour is shown in Figure 4a. It shows a surface region significantly enriched in Rh when compared to the composition of the bulk material. This region corresponds to the oxide layer and has an overall Rh composition of 37 at.% ±3.7% (calculated from the metal ions only). The end of the Rh enriched oxide layer can also be seen to correspond to the levelling

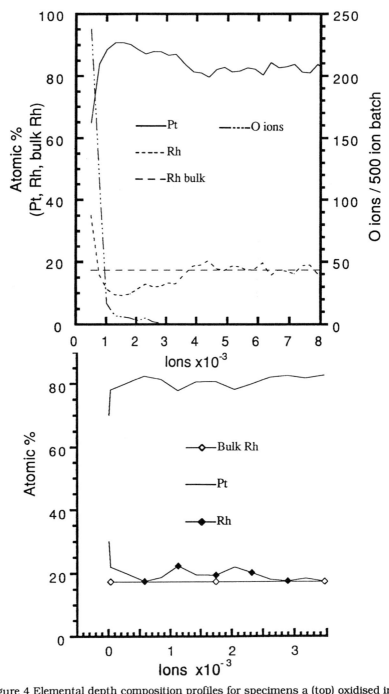

Figure 4 Elemental depth composition profiles for specimens a (top) oxidised in air at 1 atm. for 1 hour at 800°C b (bottom) oxidised in air at 1atm. for 1 hour at 800°C and then reduced in H_2 at 200°C.

out of the rate of detection of oxygen ions in the analysis. The oxide region is then followed by a Rh depleted layer in which the Rh content drops to as low as 10 at.%. The Rh depleted region corresponds to a depth of around 5 atomic planes ie 10 Å. A similar pattern was found for specimens oxidised at 600°C for 1 hour. A Rh enriched oxide layer (22 at. % Rh ±4%) was followed by a Rh depleted layer. However both the depth and extent of the Rh depletion were lower than observed for the specimen oxidised at 800°C. For the specimen oxidised at 600°C the Rh composition fell to a minimum value of 15 at.% and this slightly depleted region extended for only a monolayer (2Å).

A depth composition profile for a specimen oxidised at 800°C for 1 hour and then reduced for 30 minutes in H_2 at 200°C is shown in Figure 4b. A Rh enriched surface is found to be maintained though the composition in the surface region at 25 at.% ±4% Rh has a lower Rh enrichment than is the case for the specimen oxidised at 800°C. A Rh depleted layer is not observed.

4 DISCUSSION

It is clear from this work that oxidised and reduced specimens show clear signs of their thermal history. Carol and Mann[12] showed that there were two forms of Rh_2O_3 present in the temperature range 600-900°C. One predominating at 600-750°C the other from 750°C, the latter being a thicker oxide. This coincides with our FIM and AP observations of the oxide formed at 600°C and 800°C.

These results demonstrate that AP can give accurate absolute depth composition profiles. With the use of the related technique of FIM it is also possible to combine this data with observation of the surface structure.

REFERENCES

1. M.K. Miller and G.D.W. Smith, 'Atom Probe Microanalysis', Materials Research Society, Pittsburg, U.S.A., 1989.
2. T.T. Tsong, 'Atom-Probe Field Ion Microscopy', Cambridge University Press, Cambridge, U.K., 1990.
3. p 278 in reference 2 above.
4. K.C. Taylor, in 'Automobile Catalytic Converters in Catalysis and Automotive Pollution Control', Eds. A.Crucq, A. Frennet, Elsevier, New York,1987, Vol. 30 p.97.
5. A.D. van Langeveld, J.W. Niementsverdriet, Surf. Sci., 1986, 178, 880.
6. T.T. Tsong, D.M. Ren, M. Ahmad, Phys. Rev. B, 1988, 38, 7428.
7. F.C. M.J.M. van Delft, J. Siera, B.E. Nieuwenhuys, Surf. Sci., 1989, 208, 365.
8. A.R. McCabe, G.D.W. Smith, Plat. Met. Rev., 1988, 32, 11.
9. T. Wang, L.D. Schmidt, J. Catal., 1981, 71, 411.
10. A. Cerezo, T.J. Godfrey, G.D.W. Smith, Rev. Sci. Instrum., 1988, 59, 862.
11. A.R. McCabe, G.D.W. Smith, Plat. Met. Rev., 1983, 27, 19.
12. L.A. Carol, G.S. Mann, Oxid. Met., 1989, 34, 1.

Carbon Monoxide Hydrogenation over Alumina-supported Ruthenium–Manganese Bimetallic Catalysts

T. H. Syed*, B. H. Sakakini, and J. C. Vickerman
SURFACE ANALYSIS RESEARCH CENTRE, DEPARTMENT OF CHEMISTRY, UMIST, MANCHESTER M60 1QD, UK

Abstract
A series of ruthenium-manganese catalysts supported on alumina were prepared by coimpregnation of high surface area Al_2O_3 with $RuCl_3$ and $MnCl_2$. The catalysts were characterized by BET surface area, carbon monoxide chemisorption, Transmission Electron Microscopy (TEM), X-ray Photoelectron Spectroscopy (XPS), and Secondary Ion Mass Spectroscopy (SIMS). The catalysts were used for carbon monoxide hydrogenation reaction. Catalytic results showed that the addition of Mn while inhibiting the methanation reaction, it enhances the production of higher hydrocarbons. The influence of manganese addition on both the activity and selectivity of the CO hydrogenation is discussed in terms of the ensemble and electronic effects produced by its addition to the ruthenium catalysts.

Introduction
Introduction of bimetallic catalysts to the hydrogenation of carbon monoxide has been the subject of a wide range of investigations. The main purpose was to modify the properties of a catalyst sample that has been proved to be active in Fischer-Tropsch (FT) synthesis. Ruthenium is well known to be a very active catalyst for the hydrogenation of carbon monoxide to hydrocarbons, both methane and higher hydrocarbons (1). Manganese has been used to modify the catalytic behaviour of iron based catalysts, and in many cases it was found to improve the selectivity with regards to olefins and chain growth (2-5). The present study was undertaken to determine the effect of manganese addition on the catalytic behaviour of Ru/Al_2O_3 system in CO hydrogenation.

Experimental Section

Catalyst Preparation. The catalysts were prepared from $RuCl_3$ and $MnCl_2$ by the coimpregnation method described previously (6). All

---
[1] Present Address: Dr A. Q. Khan Research Laboratories, P. O. Box 502, Rawalpindi, Pakistan.

catalysts were reduced in flowing hydrogen at 723 K for ca. 15 h. Four samples were prepared containing ruthenium and manganese in the atomic ratios 0.00, 0.05, 0.01 and 0.20. Ruthenium loading was 1% (w/w). The final composition of the catalysts were determined by atomic absorption . The catalysts are designated as RM/ x : y where x : y is the ruthenium : manganese ratio in the sample.

Catalyst Characterization. The amount of CO adsorbed was determined at 298 K by a conventional pulse technique (6). TEM was carried out on a Philips EM400T instrument equipped with an EDAX 990-100 analytical facility. The particle size distribution was obtained by measuring approximately 400 particles. From this distribution the total metal dispersion was calculated by using methods outlined elsewhere (6). The surface concentrations of ruthenium and manganese were determined with Secondary Ion Mass Spectroscopy SIMS in its static mode. The positive secondary ions generated on bombardment of the surface of the catalyst with an argon ion beam were analyzed using a VG 12-12 quadrupole mass spectrometer. The incident argon ions had an energy of 2KeV; a current of 1nA/ cm^2 was used. XPS spectra were recorded in an ESCA-III instrument, using MgKα radiation. Targets were prepared from powder samples. Prior to the analysis the samples were sputtered with an argon ion beam with an ion current of 20 µA for 5 minutes to remove hydrocarbons adsorbed on the catalysts surface. The low atomic concentration of IB metal and the overlap of contamination C1s peak on the strongest Ru3d$_{3/2}$ signal complicate the interpretation of the spectra. Peaks used for quantitative analysis were Ru3d$_{5/2}$ and the Al$_{2p}$. Binding energies of all the elements were calculated using C1s peak as standard at the binding energy of 284 eV.

Catalytic hydrogenation of CO was carried out in a single-pass flow reactor at atmospheric pressure with a premixed gas of composition 10.45% CO : 47.40% H$_2$: 42.15% Ar (v/v/v) supplied by B.O.C. Specialist Gases Ltd. as described previously (6) . Gas chromatographic measurements of CH$_4$, CO , and various hydrocarbons was made by the same method as described previously (6). The reaction was performed in the temperature range 460-580 K.

Results and Discussion

CO Adsorption . Blank experiments on the support and on a sample of Mn dispersed on alumina revealed that under the experimental conditions CO does not adsorb on alumina or manganese. CO adsorption studies revealed a suppression in the CO uptake by the Ru-Mn/Al$_2$O$_3$ catalysts used (Fig.1). This is an indication that a substantial amount of Mn is preferentially covering Ru atoms on the surface of the catalyst. The steep decline in the CO adsorbed on the surface at low manganese

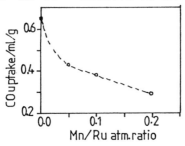

Figure 1. CO uptake/g cat. as a function of catalyst composition.

loadings could be taken as an indication that manganese is highly dispersed in this region.

TEM. The particle size distribution of two of the samples investigated by TEM is presented in Fig. 2 , while Table 1 gives the estimated average metal particle size, and total metal dispersion. The dispersion

Table 1 : Estimated Average Metal Particle Size and Metal Dispersion

Catalyst	Particle Size/nm	Metal Dispersion %
RM/100:0	4.1	23
RM/100:5	4.1	24
RM/100:10	3.8	27
RM/100:20	3.9	29

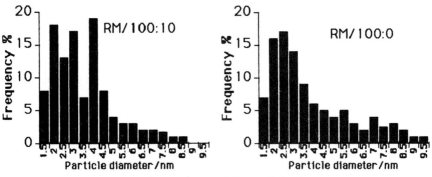

Figure 2. Particle size distribution of the catalysts.

estimated in the present study appears rather low when compared with other supported systems prepared by the impregnation method. Lai et al (6) and Xiexioan (7) have reported dispersion of around 50% for silica supported Ru-Cu . However metal dispersions of 17% and 20% have been reported for alumina supported Ru-Cu (8) and silica supported Ru-Au (8), respectively. What needs to be stressed here is that in metal supported catalysts, metal dispersion depends on many parameters, such as metal loading, metal precursor, preparation method and reduction temperature, which in many cases make comparison of metal dispersion in metal supported catalysts rather difficult . EDAX analysis revealed that particles of less than 2.5 nm were monometallic Mn, while those larger than 5.0 nm were without exception monometallic Ru. Bimetallic clusters were identified in the size range 2.5-5 nm.

SIMS. The variations of the secondary ion yield of ruthenium and manganese expressed as the intensity ratio $^{102}Ru/^{27}Al$ & $^{51}Mn/^{27}Al$ are shown in Fig. 3 as a function of catalyst composition. The Al^+ SIMS signal was taken as an internal standard to allow meaningful camparisons of surface concentration to be made from sample to sample.

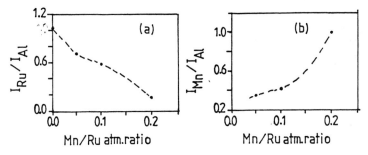

Figure 3. Plot of the secondary ion intensity ratios as a function of catalyst composition.

Fig. 3a shows a decline in the Ru^+/Al^+ intensity ratio and a simultaneous increase in Mn^+/Al^+ intensity ratio (Fig.3b) as the Mn content increases. This suggests that Mn is to some extent deposited on the ruthenium clusters of the alumina support. Similar results were obtained in ruthenium-copper catalyst supported on SiO_2 and Al_2O_3 (8,9).

XPS. XPS spectra showed a decline in the recorded intensity of Ru 3d5/2 XPS signal with increasing manganese loading, thus supporting SSIMS findings, where it was concluded that Mn is preferentially covering Ru on the surface of the catalyst. Due to the low loading of Mn metal, it was rather difficult to obtain the spectra of Mn $2p_{1/2}$ or Mn $2p_{3/2}$ below the Ru/Mn atomic ratio of 0.2.

Table 2 presents the change in binding energy and the FWHM for the Ru3d5/2 photoelectron peak with manganese loading. The Table reveals a decrease in the binding energy recorded with increasing manganese loading. Changes in metal binding energies in XPS studies on similar systems have been interpreted in terms of : (1) particle size effect, (2) electronic interaction between the metal and support or another metal and/or (3) final state effect, where lowering in the binding energy results from the screening of the core hole of the metal by electrons of

Table 2 : The Effect of Mn Addition on the B.E of Ru3d$_{5/2}$ and FWHM

Catalyst	B.E/eV	FWHM
RM/100:00	283.10	0.40
RM/100:5	282.25	0.40
RM/100:10	281.90	0.44
RM/100:20	281.60	0.50

neighbouring atoms (10,11). Although it is very difficult usually to determine which factor is the predominant one in this case, the fact that changes in the metal particle sizes as revealed by TEM in this study is not that great makes it rather difficult to envisage that it is contributing appreciably to the observed binding energy changes. It does seem however, that the changes in the binding energy shown in Table 2 could be partly due to some kind of electronic interaction between Mn and Ru, resulting in an increase in the electron density of the ruthenium. Evidence for this sort of interaction comes from two different separate studies, namely Temperature Programme Desorption of CO adsorbed on Ru-Mn/Al$_2$O$_3$ samples (12) and an EELS studies on the adsorption of CO on Ru(0001)/Mn model catalyst (13). TPD spectra revealed an increase in the area under CO$_2$ curve with a simultaneous decrease of the area under CO, suggesting an increase in the CO disproportionation with Mn addition. Taking into account that adsorption of CO on a metal occurs through donating-back-donating interactions (14) and the fact that CO adsorbs only on Ru in this case, the increase in the rate of CO disproportionation would be expected if the electron density around the ruthenium adsorption site is increased. The net result of this increase in electron density is to increase the extent of back- donation of electrons from the metal to the 2π antibonding orbital of a CO molecule, thereby increasing the strength of the carbon-metal bond, while at the same time decreasing the strength of the carbon-oxygen bond, and hence an increase in the rate of CO disproportionation would be expected on the manganese modified catalysts (15). Similar sort of conclusion as to the increase in the electron density at the Ru metal adsorption site has been reached by monitoring the change in C-O stretch frequency using Electron Energy Loss Spectroscopy (EELS) for the adsorption of CO on Ru(0001)/Mn surfaces (13).

Catalytic Properties

The significant results obtained in the investigation of the catalytic properties of the Ru-Mn/Al$_2$O$_3$ system in carbon monoxide hydrogenation are summarized in figures 4 & 5. Figure 4 shows a comparison of the catalytic activity for the production of methane and higher hydrocarbons

(C2-C4) for the different catalysts investigated at 473 K as a function of Mn loading. The hydrocarbon product distribution was measured for a wide range of temperatures. Because CO conversions can have a

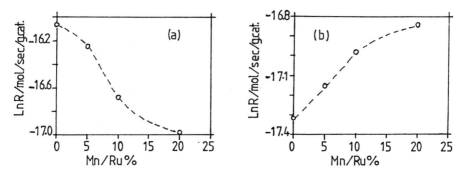

Figure 4. Catalytic activity of the Ru-Mn/Al$_2$O$_3$ catalysts in CO hydrogenation at 473 K for the production of methane (a) and higher hydrocarbons (b) as a function of catalyst composition expressed as Mn/Ru atm. ratio %.

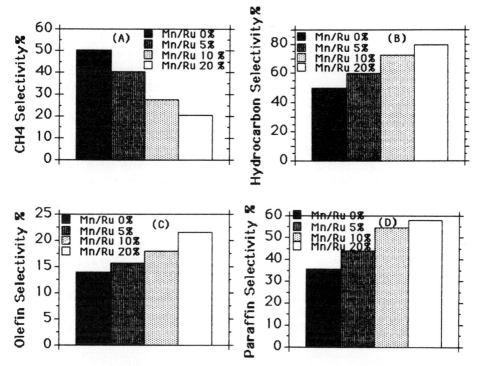

Figure 5. Comparison of product selectivities of the Ru-Mn/Al$_2$O$_3$ catalysts in CO hydrogenation at 473 K. Catalyst composition is expressed in Mn/Ru atm. ratio %.

relatively large effect on hydrocarbon product distributions (16) catalysts were compared at approximately the same conversion level. Figure 5 summarizes the selectivity for the production of methane and higher hydrocarbons (paraffins & olefins) at 473 K. Figure 4 indicate that the effect of manganese addition to the Ru/Al_2O_3 system is to reduce the methanation rate, while enhancing the rate of formation of Fischer-Tropsch products. In terms of product distribution or selectivity it is apparent that the role of manganese as a catalytic modifier is to limit methane formation in preference to hydrocarbon chain growth.

It is known that the methanation reaction is a hydrogen demanding one, hence a depression in the rate of the methanation reaction is to be expected if the availability of hydrogen on the surface of the catalyst is reduced. It has been reported that an ensemble of 4 to 6 Ru atoms is required for H_2 adsorption (17). Dalmon and Martin (18) found that on Ni-Cu alloys, the ensemble size for methanation is 12 Ni atoms. It is not surprising then to suggest, that the hydrogen adsorption would be suppressed as the result of an ensemble effect since Mn is blocking the ruthenium adsorption sites on the surface. The possibility that an electronic effect may play a role in depressing hydrogen chemisorption can also be considered in light of the previous discussion about the increase in the electron density at the ruthenium adsorption site. It is known that upon chemisorption, chemisorbed hydrogen donates electron density to the metal adsorption site (19). Increasing the electron density of the chemisorption site with manganese addition should then decrease the surface coverage of hydrogen. It is suggested that one of the effects of the manganese metal modifier is to reduce the availability of hydrogen on the surface as a result of an ensemble and/or an electronic effect, and that this leads to a suppression in the rate of CO hydrogenation to methane.

As can be seen from figure 4 the role of manganese as a catalytic modifier is not confined to limiting methane formation, but that it does this in preference to hydrocarbon chain growth. This increase may be attributed to the depression by manganese of the hydrogenation of surface intermediates which enhances chain propagation by altering the mechanistic pathways of the F-T synthesis. An increase in the number of active sites responsible for chain propagation in the catalysts investigated and whose nature is rather difficult to speculate at the moment could also be a possible explanation for the observed enhancement in hydrocarbon formation (20-22).

If one considers the observed enhancement in olefin selectivity to be at least partly due to the reduced ability of the modified catalysts to adsorb hydrogen, then the results obtained are not surprising (fig.5 C). Another reason for the enhancement of the olefin selectivity may be an acceleration in the abstraction β- hydrogen atom from the adsorbed hydrocarbon (23).

Conclusion

(1) Manganese is preferentially covering Ru on the surface of the catalysts investigated.
(2) The role of manganese as a catalytic modifier for the alumina-supported Ru catalysts prepared by coimpregnation is to suppress the methanation reaction in preference to hydrocarbon chain growth.
(3) The variation in catalytic behaviour with manganese addition to the Ru/Al_2O_3 system both in terms of activity and selectivity can be understood by considering that the role of Mn is (a) to suppress the adsorption of hydrogen both on ensemble and electronic grounds and (b) to increase the number of surface sites active in chain propagation.

References
1. M. A. Vannice, J. Catal. **37**, 449 (1975).
2. K. B. Jensen, F. E. Massoth, J. Catal. **92**, 109 (1985).
3. W. L. van Dijk, J. W. Niemandsverdriet, A. M. van der Kraan and H. S. van der Bann, Appl. Catal. **2**, 273 (1982).
4. R. Malessa and M. Baerns, Ind. Eng. Chem. Res. **27**, 279 (1988).
5. W. D. Deckwer, Y. Serpemen, M. Ralek and B. Schmidt, Ind. Eng. Chem. Proc. Des. Dev. **2**, 222 (1982).
6. S. Y. Lai and J. C. Vickerman, J. Catal. **90**, 337 (1984).
7. G. Xiexian, X. Qin, L. Yongxue, J. Dai and J. Pinliang, in " Proc. 8th Int. Cong. Catal.", Vol. **4**, 599 (1984).
8. N. I. Dunhill, Ph. D. Thesis, UMIST, 1990.
9. S. Y. Lai, Ph. D. Thesis, UMIST, 1986.
10. M. G. Mason, L. J. Gerenser and S. T. Lee, Phy. Rev. Lett. **39**,288 (1977).
11. M. Cini, Surf. Sci. **62**,148 (1977).
12. T. H. Syed, B. H. Sakakini and J. C. Vickerman, in preparation.
13. T.H. Syed, B. H. Sakakini and J. C. Vickerman, submitted for publication.
14. J. Blyholder, J. Phys. Chem. **68**,2772 (1964).
15. M. M. McClory and R. D. Gonzalez, J. Catal. **89**, 392 (1984).
16. T. Okuhara, H. Tamura and M. Misono, J. Catal. **95**, 41 (1985).
17. J. C. Vickerman and K. Christmann, Surf. Sci. **120**, 1 (1982).
18. J. A. Dalmon, G. A. Martin, " Proceedings 7th Int. Cong. Catal., Tokyo, 1980," p. 402. Kodansha/Elsevier, Tokyo/Amesterdam, 1981.
19. M. E. Dry, in ," Catalysis, Science and Technology" (J. R. Anderson and M. Boudart, Eds.), Vol. 1 Springer-Verlag, New York, 1981.
20. S. J. Madon and W. F. Taylor, J. Catal. **69**, 32 (1981).
21. L. Konig and J. Gaube, Chem. Ing. Tech. **55**, 14 (1983).
22. S. R. Morris, R. B. Moyes, P. B. Wells and R. Whyman in , "Metal-support and Metal Additive Effects in Catalysis" (B. Imelik Ed.) Elsevier, Amsterdam, 1982.
23. T. Okuhara, K. Kobayashi, T. Kimura, M. Misono and Y. Yoneda, J. C. S. Chem. Comm. p.1114 (1981).

SIMS and EELS Study of the Adsorption of CO on Mn/Ru(0001)

T. H. Syed*, B. H. Sakakini, and J. C. Vickerman
SURFACE ANALYSIS RESEARCH CENTRE, DEPARTMENT OF CHEMISTRY, UMIST, MANCHESTER M60 1QD, UK

Abstract

Static Secondary Ion Mass Spectroscopy (SSIMS) and Electron Energy Loss Spectroscopy were used to investigate the adsorption of CO on Mn/Ru(0001) surfaces under ultra-high vacuum conditions. The Mn/Ru(0001) surfaces were prepared by depositing Mn on a Ru(0001) surface at room temperature.
SSIMS studies revealed that the main function of Mn is to reduce the coverage of CO adsorbed on the surface by physically blocking the adsorption sites. SSIMS analysis of the CO-covered Mn/Ru(0001) surfaces indicates that the adsorbate molecule is predominantly linearly bonded.
EELS results, apart from confirming SSIMS findings, revealed the possible existence of an electronic interaction between the deposited Mn and Ru. This was reflected in a change in the CO stretch frequency to lower wavenumbers, with manganese deposition.

Introduction

It is well known that the addition of a second metal to a single component catalyst frequently produces a catalyst whose characteristic behaviour is very significantly different from either of the individual metal components (1).
Elucidating the reasons behind this change in catalytic performance poses a challenge to surface science. Although the ensemble (geometric) and ligand (electronic) effects have been shown to play an important role in determining the catalytic behaviour of bimetallic catalysts (2-5), there is little precision in defining their mode of operation.
In trying to understand the important parameters determining the behaviour of the supported copper-ruthenium catalytic system in carbon monoxide hydrogenation, the single crystal Cu/Ru(0001) system has proved to be a very useful model (6, 7). Striking qualitative similarities

---

[1] Present Address: Dr A. Q. Khan Research Laboratories, P.O. Box 502, Rawalpindi, Pakistan.

have been found between the two systems. The behaviour of the model single crystal Au/Ru(0001) system has also been shown to compare closely to a series of polycrystalline alumina supported bimetallic Au/Ru catalysts (8) . Despite the fact that conditions at which the "real" and the 'model" catalysts are studied are rather different (9), many other researchers have proved the utility of model single crystal systems in trying to elucidate the behaviour of supported metal catalysts. Manganese has been used as a modifier for the iron based Fischer-Tropsch (FT) catalysts by many research groups (10-15) . In many cases manganese addition resulted in an improvement in the selectivity with regard to olefin formation (10-14). In a recent study on alumina supported bimetallic Mn/Ru catalysts used in CO hydrogenation, we have found that Mn suppresses the methanation reaction in preference to hydrocarbon chain growth. It also enhances the formation of olefins (16). As part of a continuing programme of research into the surface reactivity of Ru based bimetallic model catalysts we report here an investigation of the effect of surface Mn on the adsorption of CO on Ru(0001).

Experimental

The experiments were carried out in a UHV system described previously (17). The cleaning procedure of the Ru(0001) single crystal was the same as that used in other studies (18). Mn/Ru(0001) bimetallic surfaces were prepared by evaporating Mn onto the Ru(0001) surface held at 300 K. Since carbon monoxide was found to adsorb onto Ru but not onto manganese, the relative areas of the two TPD curves recorded for the Ru(0001) and each Mn/Ru(0001) surface, after exposing each of the two surfaces to the saturation coverage of CO at 300 K (15 L) were used to obtain an approximate coverage of manganese on the surface.

After the preparation of the required surface, it was exposed successively to carbon monoxide starting from 0.1L up to 15 L (1L = 10^{-6} Torr sec) and SSIMS or EELS spectra were recorded. The SSIMS studies were carried out using 0.5 nA cm^{-2} and 2 KeV Argon-ion beam, which results in negligible damage during the time-scale of the experiment (usually 5-10 mins). The EELS experiments were carried out using a 5 eV electron beam having an energy width of ca. 7meV. The adsorption of CO was studied using EELS and SSIMS at 300 K.

Results and Discussion

1. Static Secondary-ion Mass Spectrometry

Figure 1 shows a typical low resolution SSIMS spectra recorded after 15 L CO exposure for a Mn/Ru(0001) surface with θ_{Mn}=0.2 ML . It can be concluded that CO adsorbs only on Ru sites under the conditions in which the investigation was conducted. No molecular or dissociative adsorption of CO occurs on surface Mn. This is confirmed by the absence of any MnCO$^+$ signal for molecular adsorption as well as MnC$^+$ and MnO$^+$ signals indicative of dissociative adsorption.

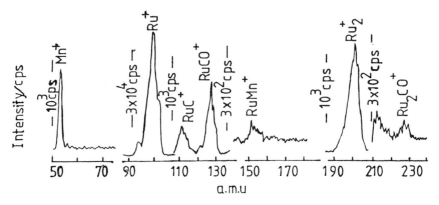

Fig.1. SSIMS spectra of CO adsorption at 300 K on Mn/Ru(0001) surface with manganese coverage of 0.2 ML.

Previously Brown and Vickerman (19) showed that the sum of Ru$_x$CO$^+$/Ru$_x^+$ can be used as a measure of the CO coverage on the surface. In figure 2, Σ Ru$_x$CO$^+$/Ru$_x^+$ is plotted as a function of manganese coverage after 15 L CO exposure . Figure 2 (a) shows that the carbon monoxide adsorption is being suppressed by the addition of manganese on the surface. This suggests that manganese blocks the Ru adsorption sites. Previous entensive investigations have demonstrated that CO adsorbate site geometry can be monitored from the SSIMS fragmentation pattern (20-22). Thus the relative yield of MCO$^+$,M$_2$CO$^+$ and M$_3$CO$^+$ ions indicates whether linear,bridge or triply bridge CO is present on the surface. Table 1 shows the relationship. The variation of the ratio MCO$^+$/ ΣM$_x$CO$^+$ is a useful parameter to monitor the population of linear CO. For the adsorption of CO on Mn/Ru(0001) the MCO$^+$/ Σ M$_x$CO$^+$ ratio presented in figure 2 (b) is nearly constant at 0.9 as a function of Mn MCO$^+$/ ΣM$_x$CO$^+$ is a useful parameter to monitor the population of linear CO. For the adsorption of CO on Mn/Ru(0001) the MCO$^+$/ Σ M$_x$CO$^+$ ratio presented in figure 2 (b) is nearly constant at 0.9 as a function of Mn coverage indicating linear coordination at all coverages,similar to that observed on a Ru(0001) surface. This is a clear indication that the coordination of CO to the surface at 300 K is not affected by the presence of Mn on the surface. This result is similar to that reached for Au/Ru(0001) bimetallic system (18). No RuMnCO$^+$ signal was observed in the SSIMS recorded spectra which indicates the absence of 'mixed' site bonding something which was found in the case of the Cu/Ru(0001) system (18).

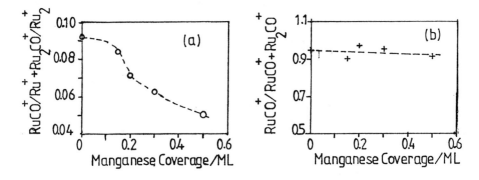

Fig. 2. Variation of (a) RuCO$^+$/Ru + Ru$_2$CO$^+$/Ru$_2$ and (b) RuCO$^+$/(RuCO$^+$ +Ru$_2$CO$^+$) ions as a function of manganese coverage.

Table 1. SSIMS 'fragmentation' pattern for different CO adsorbate structures.

	proportion of M$_x$CO ions		
adsorbate structure	MCO$^+$	M$_2$CO$^+$	M$_3$CO$^+$
linear 'on top' CO	0.9	0.1	–
two-fold bridge CO	0.3	0.6	0.1
three-fold bridge CO	0.3	0.4	0.3

2. Electron Energy Loss Spectroscopy

Figure 3 summarizes the recorded EELS spectra on the different Mn/Ru(0001) surfaces prepared after 15 L CO exposure at 300 K. The spectra shows a clear continual decrease in the relative intensity of the CO stretch band (around 250 meV) with the increase in Mn coverage. This is an indication of the decline in the amount of CO adsorbed as Mn is progressively added to the Ru(0001) surface, confirming the SSIMS observation. Whilst this was attributed to the physical blocking of the Ru adsortpion sites the EELS spectra reveal a continual decrease in the frequency of the CO stretching vibration with Mn addition. Table 2 gives

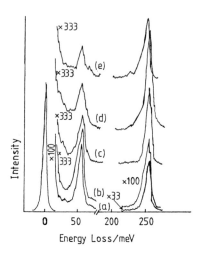

Fig.3. EELS spectra of CO adsorption (15 L exposure) at 300 K on (a) clean Ru(0001) surface and Mn/Ru(0001) surfaces having Mn coverages of (b) 0.06, (c) 0.10, (d) 0.30 and (e) 0.50.

Table 2. CO stretching vibration frequencies in cm-1 observed for the Mn/Ru(0001) system with varying manganese coverages.

Mn coverage	Frequency/cm^{-1}
0.00	2049 (254)
0.06	2033 (252)
0.10	2033 (252)
0.30	2025 (251)
0.50	1988 (246.5)

the CO stretching frequency in cm^{-1} for the different manganese coverages. The number in parenthesis is the frequency in meV (1meV=8.066 cm^{-1}).

Band shifts in vibrational spectroscopic studies in the adsorption of CO on supported metal catalysts and single crystals have been attributed mainly to three different influences, namely: (i) the electronic state of the metal on which a CO molecule is adsorbed(23-24); (ii) dipole-dipole interaction between the adsorbed CO molecules (25), and/or (iii) particle size (26). The last effect can be ignored in our present study, since adsorption is taking place on the Ru single crystal and not on the deposited Mn.

The influence of electronic factor resides in the fact that the adsorption of CO on a metal occurs through an electron donation-back-donation process (27). The decrease in the CO stretching vibration with the Mn addition could be explained by the presence of an electronic effect brought about by an increase in the electron density at the Ru adsorption site. The net result of such an increase in electron density would be to increase the extent of back-donation, hence increasing the strength of the carbon-metal bond, while at the same time decreasing the strength of the carbon-oxygen bond, causing the CO stretching vibration to decrease as observed. Thus, the role of the Mn added may be to increase the electron density of the Ru adsorption site, and enhance the back-donation of electrons from the Ru metal to the anti-bonding orbital of the adsorbed CO molecule.

However, although this interpretation for the red shift (downshift) seems to be reasonable, a similar red shift could be caused by a variation in the mutual interaction of the CO dipoles, in this case the decoupling of CO dipole-dipole interaction. This would be caused by the disruption in the periodicity of the adlayer induced by the presence of the manganese metal adatoms which would increase the CO-CO distance as the amount of CO adsorbed on the surface decreases.

Further studies are needed to distinguish the electronic (i.e., chemical bonding) effects from "geometric" (i.e., coupling) effect. This has proved to be possible using the isotopic dilution method in which the dipole-dipole coupling effect is dramatically reduced to the minimum if not eliminated completely. Once the coupling effect has been eliminated a more reliable discussion of the electronic effect is possible.

Conclusion

1. SSIMS revealed that no associative or dissociative adsorption of CO occurs on deposited Mn at 300 K. Mn merely blocks Ru adsorption sites on the surface. This is clearly indicated by the reduction in the amount of CO adsorbed on the surface. Mn does not influence the adsorbate site geometry.

2. EELS results apart from confirming the findings of SSIMS revealed the possible existence of an electronic interaction between the deposited Mn and Ru.

3. The study reveales the complementary nature of SSIMS and EELS techniques in monitoring adsorbates on metal surfaces.

References

1. J. H. Sinfelt, Acc. Chem. Res. **10**, 15 (1977).
2. G. C. Bond and B. D. Turnham, J. Catal. **45**, 128 (1976).
3. S. Galvango and G. Parravano, J. Catal. **57**, 272 (1979).
4. C. Betizeau, G. Leclercq, R. Maurel, C. Bolivar, H. Charcosset, R. Frety and L. Tournayan, J. Catal. **45**, 179 (1976).

5. H. C. De Jongste, V. Ponec and F. G. Gault, J. Catal. **63**, 395 (1980)
6. J. C. Vickerman and K. Christmann, Surf. Sci. **120**, 1 (1982).
7. S. Y. Lai and J. C. Vickerman, J. Catal. **90**, 337 (1984)
8. N. I. Dunhill, B. Sakakini and J. C. Vickerman,
9. D. W. Goodman and J. E. Houston, Science 236, 403 (1987).
10. R. Malessa and M. Baerns, Ind. Eng. Chem. Res., **27**, 279 (1988).
11. W. Deckwer, Y. Serpemen, M. Ralek and B. Schmidt, Ind. Eng. Chem. Process Des. Dev. **21**, 222 (1982).
12. J. Venter, M. Kaminsky, G. L. Geoffroy and M. A. Vannice, J. Catal. **103**, 450 (1987).
13. L.I. Kuznetsova, K. G. Nguen, A. R. Suzdrof, L. A. Beilin, E. S. Shpiro and Kh. M. Minachev, Kinet. Katal. **40**, 830 (1989) (Eng. Transl.)
14. T. Grzybek, H. Papp and M. Baerns, Appl. Catal. **29**, 335 (1987)
15. W. L. van Dijk, J. W. Niemantsverdriet, A. M. van der Kraan, and H. S. van der Baan, Appl. Catal. **2**, 273 (1982).
16. S. T. Hussain, B. Sakakini and J. C. Vickerman, in preparation.
17. C. Harendt, B. Sakakini, J. A. van den Berg and J. C. Vickerman, J. Elec. Spec. Rel. Phen. **39**, 35 (1986).
18. B. Sakakini, A. J. Swift, J. C. Vickerman, C. Harendt and K. Christmann, J. Chem. Soc., Faraday Trans. 1, **83**, 1975 (1987).
19. A. Brown and J. C. Vickerman, Surf. Sci., **117**, 154 (1982).
20. A. Brown and J. C. Vickerman, Surf. Sci., **151**, 319 (1985).
21. A. Brown and J. C. Vickerman, Surf. Sci., **47**, 384 (1975).
22. A. Brown and J. C. Vickerman, Surf. Sci., **124**, 267 (1983).
23. G. A. Somorjai, J. Phy. Chem. **86**, 310 (1982).
24. M. M. McClory, R. D. Gonzalez, **89**, 392 (1984).
25. A. Crossley, D. King, Surf. Sci. **95**, 131 (1980).
26. F. J. C. M. Toolenaar, A. J. C. M. Bastein, and V. J. Ponec, J. Catal. **82**, 35 (1983).
27. J. Blyholder, J. Phys. Chem. **68**, 2772 (1964).

An Electron Microscopy Study of the Sintering of Pt/Rh/Alumina/Ceria Catalysts

A. Cook, A. G. Fitzgerald, and J. A. Cairns
DEPARTMENT OF APPLIED PHYSICS AND ELECTRONIC &
MANUFACTURING ENGINEERING, UNIVERSITY OF DUNDEE,
DUNDEE DD1 4HN, UK

ABSTRACT

Electron microscopy is being used to study the thermal sintering behaviour of existing and new materials for catalytic conversion in vehicle exhausts. Pt/alumina, Pt/ceria, Pt/alumina/ceria, Rh/alumina, Rh/ceria, Rh/alumina/ceria, and Pt/Rh/alumina/ceria catalysts have been studied using transmission electron microscopy (TEM), scanning electron microscopy (SEM) and energy dispersive x-ray microanalysis (EDX). The specimens were prepared by a method based on the sol-gel technique and subjected to heat treatments of 1, 2, 4, 8, 16 and 24hrs in air at 600°C, 700°C and 800°C. The results suggest that as the length of heat treatment increases and as the temperature increases, the amount of sintering increases significantly for those specimens with an alumina support. In the specimens heated for 16hrs or longer unusually large particles can be observed. EDX analysis indicates that these particles are composed of Pt. The effect of ceria in the support for the catalyst appears to be to reduce the sintering. No large Pt particles or clusters of particles were observed in any of the specimens supported by ceria, and considerably less sintering was observed for the specimens with a 5:1 ratio of alumina:ceria as a support. For comparison, a commercial vehicle exhaust catalytic converter was cut into sections and subjected to heat treatments of 6, 16 and 24 hours at 600°C, 700°C and 800°C prior to examination by SEM. Some sintering was observed and the EDX results suggest that the Pt is associated with the Ce. These results are important in that they show that the migration of the precious metal over the support can be inhibited by interposing a different material between it and the support, thereby extending the efficiency and life of the catalyst.

1 INTRODUCTION

The thermal sintering, in air, of supported metal catalysts is an important factor in the deactivation of the catalyst because of the associated loss of exposed surface area both of the metal and of the support. This is especially important in the case of vehicle exhaust catalysts because they often operate at relatively high temperatures (typically 500-900°C). Such catalysts usually contain Pt and Rh supported on alumina. In addition, so-called 3-way vehicle exhaust catalysts contain ceria in various proportions. The sintering behaviour of supported Pt particles has been studied extensively using a variety of techniques, such as transmission electron microscopy (TEM)[1-4,8-11], temperature programmed desorption spectroscopy[5,6], and extended x-ray absorption fine structure studies[7].

Transmission electron microscopy gives information on both the catalyst particle size and distribution. In this investigation TEM has been used to study the thermal sintering behaviour of model Pt, Rh and Pt/Rh catalysts supported by alumina, ceria and a combination of alumina and ceria. The results obtained from TEM have been compared with the results of a scanning electron microscopy investigation of the thermal sintering of the catalyst particles of a commercial vehicle exhaust catalytic converter.

2 EXPERIMENTAL

Thin specimens of Pt/alumina, Pt/ceria, Pt/alumina/ceria, Rh/alumina, Rh/ceria, Rh/alumina/ceria, and Pt/Rh/alumina/ceria were prepared for transmission electron microscopy by the following method. For the 2%Pt/ceria specimen 0.067g tetrammine platinous chloride was dissolved in 4ml ceria sol containing $450gl^{-1}$ ceria plus 1ml distilled water. For the 2%Pt/alumina/ceria specimen 0.056g tetrammine platinous chloride was dissolved in 0.55ml of the ceria sol plus 5ml alumina sol containing $250gl^{-1}$ alumina to give a 5:1 ratio of alumina:ceria. A 2%Pt/alumina specimen was prepared by a similar method. The specimens containing rhodium were prepared using rhodium trichloride dissolved in alumina or ceria sol, or a mixture of both to give 2%Rh in alumina, ceria, or a 4:1 ratio by weight of alumina:ceria. A neutral detergent was added to each solution as a wetting agent. Stainless steel TEM specimen grids were dipped into the solutions, dried in air at 80°C for 30 minutes and fired in air at 600°C for 30 minutes. This resulted in thin sheets of the catalyst being suspended across the corners of some of the grid squares. The specimens were then subjected to heat treatments of 1, 2, 4, 8, 16 and 24 hours in air at 600°C, 700°C and 800°C.

For comparison, a commercial vehicle exhaust catalytic converter was prepared for study by scanning electron microscopy by cutting it into sections and firing it in air for 6, 16 and 24 hours at 600°C, 700°C and 800°C. Prior to examination these specimens were cut again to provide a fresh surface for study.

For transmission electron microscopy a JEOL 100C STEM was operated at 100kV in bright field mode at either 100,000x or 160,000x depending on the stability and charging of the specimens. Micrographs were obtained from several areas of each sample to give an accurate account of the effect of the heat treatment.

A JEOL T300 scanning electron microscope operated at 15kV was used to obtain backscattered electron images (BEI), with atomic number contrast, of the surface of the commercial vehicle exhaust catalytic converter. Specimen charging restricted the magnification used to 3,500x and below.

Energy-dispersive x-ray microanalysis (EDX) could be carried out in either microscope using a Link Systems x-ray detector interfaced to a DAPPLE system. The area of the specimen analysed by EDX is around $0.5\mu m$.

3 RESULTS

Figure 1 shows the Pt/alumina specimens after heat treatments of 1hr at 600°C, 16hrs at 700°C and 4hrs at 700°C. There appears to be an increase in the size of the Pt particles even after only a few hours heating in air at 600°C and many areas such as that shown in figure 1(c) where unusually large Pt particles occur can be observed. The dark areas on the specimens were shown by EDX analysis to be composed of Pt.

Figure 2 shows the Pt/ceria specimens after heat treatments of 2hrs at 600°C and 8hrs at 800°C. The Pt particle size does not appear to increase significantly relative to the background support particles and the particle distribution appears to remain stable even after heat treatment of 24hrs at 800°C although the support particle size increases on heating. No abnormally large Pt particles could be seen on any of the Pt/ceria specimens.

Figure 3 shows the Pt/alumina/ceria specimens, with an alumina:ceria ratio of 5:1 by weight, after heat treatments of 4hrs at 600°C and 24hrs at 800°C. Some thermal sintering of the Pt particles can be observed in these specimens but the effect is reduced when compared with the sintering observed in the specimens with an alumina - only support. No abnormally large Pt particles can be seen on any of the Pt/alumina/ceria specimens.

Figure 4 shows the Rh/alumina specimens after heat treatments of 8hrs at 600°C and 16hrs at 800°C. Some sintering of the Rh particles can be seen but no unusually large Rh particles are visible.

Figure 5 shows the Rh/ceria specimens after heat treatments of 2hrs at 600°C and 24hrs at 800°C. There is a dramatic increase in the support particle size although this increase is not observed in the Rh particles. No areas with unusually large Rh particles or clusters of particles can be seen on any of the specimens supported by ceria.

Figure 6 shows the Rh/alumina/ceria specimens, with an alumina:ceria ratio of 4:1 by weight, after heat treatments of 2hrs at 600°C and 16hrs at 800°C. There is a slight increase in the size of the support particles on heating, which could be due to the higher proportion of ceria in the support compared with the Pt/alumina/ceria specimens described above. Some sintering of the Rh is observed, but the effect appears to be much less than for the Rh/alumina specimens.

Figure 7 shows the Pt/Rh/alumina/ceria specimens, with an alumina:ceria ratio of 4:1 by weight, after heat treatments of 8hrs at 600°C and 16hrs at 800°C. Again there is a slight increase in the support particle size but very little sintering of the catalyst can be seen on any of the specimens supported by both alumina and ceria.

Figure 8 shows backscattered electron images of the surface of the commercial vehicle exhaust catalytic converter taken in the SEM after heat treatments of 6hrs at 600°C and 16hrs at 700°C. EDX analysis shows the bright areas on the images to be composed of Pt and Ce. Neither element could be detected in significant quantities without the other. Some thermal sintering of the catalyst particles occurs, although this effect is not as marked as the sintering of the TEM specimens.

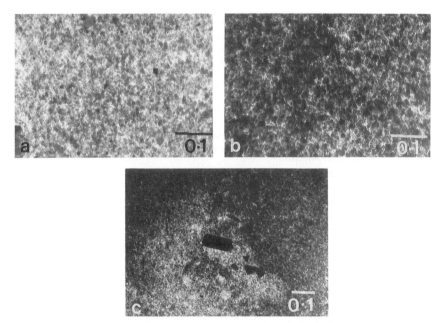

Figure 1. Pt/alumina specimen after heat treatment of (a) 1hr 600°C, (b) 16hr 700°C and (c) 4hr 700°C. Scale bars in μm.

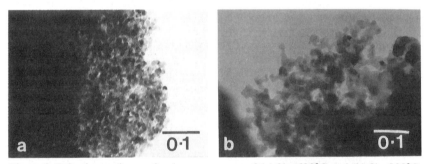

Figure 2. Pt/ceria specimen after heat treatment of (a) 2hr 600°C and (b) 8hr 800°C. Scale bars in μm.

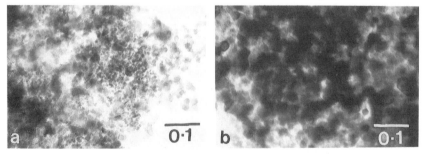

Figure 3. Pt/alumina/ceria specimen after heat treatment of (a) 4hr 600°C and (b) 24hr 800°C. Scale bars in μm.

The Sintering of Pt/Rh/Alumina/Ceria Catalysts

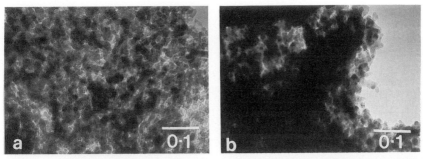

Figure 4. Rh/alumina specimen after heat treatment of (a) 8hr 600°C and (b) 16hr 800°C. Scale bars in μm.

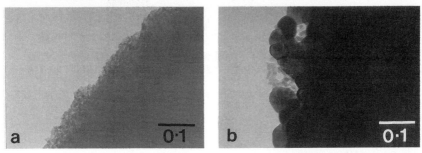

Figure 5. Rh/ceria specimen after heat treatment of (a) 2hr 600°C and (b) 24hr 800°C. Scale bars in μm.

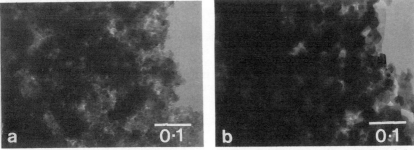

Figure 6. Rh/alumina/ceria specimen after heat treatment of (a) 2hr 600°C and (b) 16hr 800°C. Scale bars in μm.

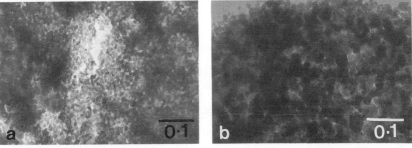

Figure 7. Pt/Rh/alumina/ceria specimen after heat treatment of (a) 8hr 600°C and (b) 16hr 800°C. Scale bars in μm.

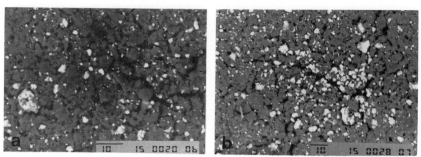

Figure 8. Backscattered electron images of commercial vehicle exhaust catalytic converter after heat treatment of (a) 6hr 600°C and (b) 16hr 700°C. Scale bar 10μm.

4. DISCUSSION

From TEM investigations it would appear that thermal sintering of the Pt particles occurs when an alumina-only support is used, but not if ceria is used as the support. When a proportion of ceria is added to the alumina support the amount of sintering is greatly reduced. This would suggest that the Pt is associated with the ceria. EDX analysis results from the TEM specimens and from the commercial catalytic converter surface support this view since on samples containing both Pt and Ce neither can be identified separately.

It would appear that the ceria support undergoes a change when subjected to heat treatments of more than a few hours at the temperatures used in this work. This change is indicated by the large increase in the size of the cerium oxide support particles and is most noticeable at the highest temperature used, 800°C.

In the specimens where sintering is a major effect it appears that particle migration and coalescence are important since few small catalyst particles can be seen on the specimens which have undergone more severe heat treatments. This is consistent with previous work[1] although it is possible that small particles exist which are below the size observable with the microscope used.

References

1 Harris,P.J.F.; Boyes,E.D.; Cairns,J.A.   J. Catal. vol 82 (127-146) 1983
2 White,D.; Baird,T.; Fryer,J.R.; Freeman,L.A.; Smith,D.J.; Day,M.   J.Catal. vol 81 (119-130) 1983
3 Sushumna,I,; Ruckenstein,E.   J.Catal. vol 109 (433-462) 1988
4 Bellare,A.; Dadyburjor,D.B.; Kelley,M.J.   J.Catal. vol 117 (78-90) 1989
5 Beck,D.D.; Carr,C.J.   J.Catal. vol 110 (285-297) 1988
6 Altman,E.I.; Gorte,R.J.   J.Catal. vol 110 (191-196) 1988
7 Bazin,D.; Dexpert,H.; Lagarde,P.; Bournonville,J.P.   J.Catal. vol 110 (209-215) 1988
8 Baker,R.T.K.; Prestridge,E.B.; Garten,R.L.   J.Catal. vol 56 (390-406) 1979
9 Baker,R.T.K.; Thomas,C.; Thomas,R.B.   J.Catal. vol 38 (510-513) 1975
10 Flynn,P.C.; Wanke,S.E.   J.Catal. vol 37 (432-448) 1975
11 Chu,Y.F.; Ruckenstein,E.   J.Catal. vol 55 (281-298) 1978

EXAFS Characterization of Surface Species in Na-doped Molybdena–Titania Systems

P. Malet[1], A. Muñoz-Paez[1], C. Martin[2], I. Martin[2], and V. Rives[2]

[1] DEPARTAMENTO DE QUIMICA INORGÁNICA, UNIVERSIDAD DE SEVILLA, FACULTAD DE QUIMICA, 41012 SEVILLA, SPAIN
[2] DEPARTAMENTO DE QUIMICA INORGÁNICA, UNIVERSIDAD DE SALAMANCA, FACULTAD DE FARMACIA, 37007 SALAMANCA, SPAIN

1 INTRODUCTION

Supported molybdenum oxide is one of the most classical and important examples of solid catalysts because of its high activity in industrial processes. The activity and selectivity of these systems is deeply modified by the structure and physicochemical properties of the dispersed moieties supported on the surface and therefore can be affected by parameters such as preparation conditions, nature of the support, active phase contents or addition of dopants. When TiO_2 is used as support it can act as a promoting agent of the active phase and the systems so obtained are effective in the selective oxidations of propene[1], butene[2] and methanol[3]. Also the use of alkali metals as promoters has been described when MoO_3 is supported on alumina and changes in the coordination of Mo species from octahedral to tetrahedral species induced by the presence of these dopants have been proposed on the basis of changes in the reducibility of the supported phase[4] or Raman and ultraviolet spectroscopies[5]. In the present contribution a set of monolayer MoO_3 catalysts supported on TiO_2 and doped with different amounts (1 or 3% w/w) of sodium have been characterized by means of EXAFS spectroscopy.

2 EXPERIMENTAL

Supported samples were prepared by impregnation methods. Different amounts of sodium (1 or 3% by weight) were incorporated to TiO_2 (Degussa P-25) from NaOH solutions. After calcination in air at 773 K for 3 h the supports were impregnated with an aqueous solution of ammonium heptamolybdate (from Carlo Erba), by adding the amount needed to achieve one theoretical monolayer of MoO_3 (5.9% Mo w/w), as calculated from the BET surface area of the support (53±2 m²/g) and the area occupied by a "molecule" of MoO_3 (15x10⁴ pm²). The impregnated products were dried at 383 K for 24 h and calcined at 773 K for 3 h in air.

X-ray absorption data were collected at 77 K on wiggler station 9.2 at the Synchrotron Radiation Source (SRS) in Daresbury (England), with a beam of 2GeV, 210-230 mA and with 50% higher harmonic rejection. Samples were ground and homogenized, and pre-calculated amounts were pressed into self-supporting wafers with a mass absorption coefficient μx 2.5 cm^{-1} just above the Mo K edge (20005 eV). Reference compounds (MoO$_3$ and Na$_2$MoO$_4$.2H$_2$O) were diluted with boron nitride. At least three scans were recorded and averaged in order to obtain the experimental spectra. The EXAFS function (χ) was obtained from the experimental X-ray absorption spectrum by subtracting a quadratic expansion based on the extrapolation of the pre-edge region, followed by a cubic spline background removal. The resulting EXAFS function was normalized to the height of the absorption edge and analyzed for structural information by fitting with k^1 and k^3 weighting schemes in both k- and r-spaces.

3 RESULTS AND DISCUSSION

Fig.1 includes uncorrected k-weighted Fourier Transforms (FT) of EXAFS spectra for two reference compounds, crystalline MoO$_3$ and Na$_2$MoO$_4$.2H$_2$O (Fig. 1a), and for the Na-doped MoO$_3$/TiO$_2$ samples (Fig. 1b).

The radial distribution function derived from the EXAFS of sodium molybdate contains only a main peak (1.25 Å, uncorrected) that has been ascribed[6,7] to four oxygen neighbours at 1.772 Å, according to its structure as determined by X-ray diffraction[8], and the existence of tetrahedral MoO$_4$ species in this compound.

EXAFS oscillations for MoO$_3$ are much more complicated, with peaks centered at 1.5 and 3.1 Å in the uncorrected FT, Fig. 1a. The first maximum has been ascribed to Mo-O contributions[6] in the distorted MoO$_6$ octahedrons that built the MoO$_3$ structure, while the peak at 3.1 Å is mainly due to Mo-Mo absorber-backscatterer pairs[6].

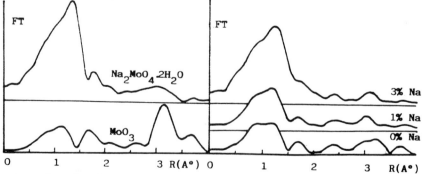

Fig. 1. Absolute part of the uncorrected k-weighted FT of the EXAFS spectra of (a) reference compounds and (b) MoO$_3$/TiO$_2$/Na samples (data range: 3.6<k<15.0 Å^{-1}).

Comparison of FT's obtained for supported samples with those of reference compounds (Fig. 1) shows that the regular structure at long range present in MoO_3 (peak at 3 Å) is vanishing in the supported samples, indicating smaller sized MoO_3 particles, although the local order of oxygen ions around the Mo ions is maintained, leading to a complex maximum centered at 1.5 Å, intermediate between those of Na_2MoO_4 and MoO_3, thus suggesting that the local order around the Mo ions in the supported samples is intermediate between the limiting situations existing in the reference compounds.

EXAFS analysis of these Mo-O first shell contributions has been performed by isolating them by inverse FT in the 0.5-2 Å range. Experimental backscattering amplitude and phase functions for the Mo-O absorber-backscatterer pair were obtained from the Na_2MoO_4 spectrum (forward FT: k^3-weighted 3.6<k<16 Å^{-1}; inverse FT: 0.3<R<2.5 Å; four oxygen neighbours at 1.772 Å). Mo-O first-shell contributions in MoO_3 (Fig.2) can be simulated by equal number (N) of oxygen atoms at three different distances (R), in excellent agreement with crystallographic data (oxygen atoms at 1.671, 1.734, 1.948 (x2), 2.251 and 2.332 Å[9]) and yielding a total coordination number of 6.0 O/Mo up to 2.3 Å. Next nearest neighbours are at distances larger than 3.4 Å[9] and therefore are well resolved from this first shell contributions.

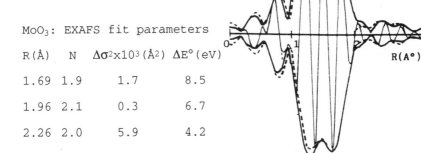

MoO_3: EXAFS fit parameters

R(Å)	N	$\Delta\sigma^2 \times 10^3$ (Å^2)	$\Delta E°$ (eV)
1.69	1.9	1.7	8.5
1.96	2.1	0.3	6.7
2.26	2.0	5.9	4.2

Fig. 2. Uncorrected k^3-weighted FT (4<k<13.5 Å^{-1}) of the isolated Mo-O first shell contributions for MoO_3 (solid line) and best-fit function (dashed line).

The approximation to the MoO_6 distorted octahedron in bulk MoO_3 obtained in this fit is similar to that used by Mensch et al.[6], who reported a total coordination number of 5.3 O/Mo up to 2.3 Å by using three different Mo-O distances to fit the EXAFS spectrum of this compound. Recently, Kisfaludi et al.[7] introduced more Mo-O first shell distances up to 2.33 Å to fit the EXAFS spectrum of a non-calcined MoO_3/Al_2O_3 physical mixture, that they claimed only contains pure MoO_3. However, the use of a more complicated model does not lead to an improvement in the fit of the FT in the low R region and yields a total coordination number of 3.9 O/Mo up to 2.33 Å, in poor agreement with crystallographic data. Therefore, this model has been discarded in our analysis.

Data in Fig. 3a correspond to first-shell contributions for the samples, as compared with those for the known structures of the reference compounds (distorted MoO_6 octahedra in MoO_3 and regular MoO_4 tetrahedra in Na_2MoO_4, Mo(Oh) and Mo(Td), respectively), and suggest that the local structure around the Mo atoms in these samples is intermediate between the limiting situations existing in the reference compounds. To assess this first guess, theoretical EXAFS spectra for different mixtures of Mo(Td) and distorted Mo(Oh) coordinations have been calculated (Fig. 3b), employing the set of parameters previously determined for MoO_3 and $Na_2MoO_4 \cdot 2H_2O$. Comparison of experimental and theoretical spectra in Fig. 3 demonstrates that changes in the EXAFS spectra could be explained by assuming an increase in the fraction of Mo(Td) in the supported samples as the Na content increases. Therefore, four Mo-O contributions should be employed to best fit this first shell: three to model the distorted octahedrons and the last one for the regular tetrahedrons.

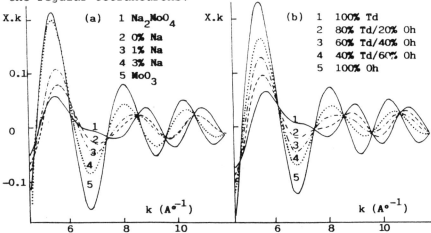

Fig. 3 (a) Isolated Mo-O first-shell contribution for $MoO_3/TiO_2/Na$ samples and reference compounds; (b) theoretical EXAFS spectra assuming mixtures of MoO_6 and MoO_4 units.

Since with the analysis ranges here employed (Δk= 9.5 Å^{-1}, ΔR= 2 Å) only three shells are statistically allowed by the Brillouin equation (maximum number of free parameters = $2 \cdot \Delta R \cdot \Delta k / \pi$), some restrictions have been superimposed to the analysis in order to ensure the physical meaning of the result. Thus, the Mo-O distances are kept equal to those present in the crystalline compounds. Moreover, since the EXAFS oscillations are normalized to one molybdenum atom, if only Mo(Oh) and Mo(Td) entities are present, the following condition should be fulfilled within the limits of precision in coordination numbers given by the EXAFS technique:

$N(Oh)/6 + N(Td)/4 = 1$

N(Oh) being the oxygen coordination number at distances characteristic of the MoO_6 octahedron and N(Td) the oxygen coordination number at the Mo-O distance in MoO_4 tetrahedra.

Table 1: Results of EXAFS analysis for $MoO_3/TiO_2/Na$ samples.

%Na	R(A)	N(Oh)	N(Td)	$\Delta\sigma^2 \times 10^3$	$\Delta E°$ (eV)	% Td
0	1.69	1.5	-	1.3	6.0	
	1.96	1.6	-	2.6	8.1	
	2.26	1.9	-	10.0	2.3	
	1.77	-	0.5	-2.2	5.5	12
1	1.69	1.3	-	0.3	4.7	
	1.96	1.7	-	3.4	8.8	
	2.26	1.9	-	8.5	0.6	
	1.77	-	1.3	1.1	4.2	29
3	1.69	0.5	-	1.2	1.9	
	1.96	0.5	-	-0.8	3.5	
	2.26	0.5	-	8.5	8.4	
	1.77	-	3.6	2.7	0.8	78

Best fit results are plotted in Fig. 4 and the calculated parameters are included in Table 1. The model provides an excellent fit of the experimental EXAFS spectra, and is coherent with the presence on the support surface of MoO_6 and MoO_4 moieties only (number of Mo atoms 1.07±0.07) with the same Mo-O distances as in the bulk compounds. A small fraction (12%) of Mo atoms in tetrahedral coordination are already present in the Na-free sample, this value increasing up to 78% as Na is added up to an atomic ratio of Na/Mo=2.9. This result is in agreement with those by Kantschewa et al.[5], who reported that almost all Mo atoms are tetrahedrally coordinated in a $NiMo/Al_2O_3$ catalyst when the sample is doped with K_2CO_3 (K/Mo atomic ratio = 2.5), and O'Young et al.[4], who also proposed that Mo ions supported on Al_2O_3 are tetrahedrally coordinated when the alkaline/Mo atomic ratios are larger than two both in K- and Cs-doped samples. The special level of interaction proposed by the latter authors when the Alkaline/Mo ratios are higher than two is not confirmed by the EXAFS study here reported, since increasing the sodium content from

Na/Mo=0 to 2.2 steadily increases the number tetrahedrally coordinated Mo atoms, thus suggesting the reaction between sodium and molybdenum atoms on the surface of the TiO_2 support to yield sodium molybdate species.

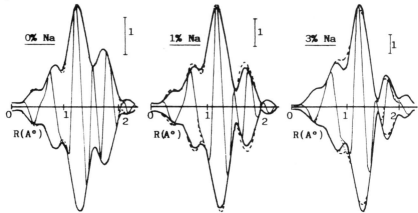

Fig. 4. Uncorrected k^3-weighted FT (4<k<13.5 Å^{-1}) of the isolated Mo-O first shell contributions for MoO_3/TiO_2/Na supported samples (solid line) and best-fit functions (dashed line).

4 ACKNOWLEDGEMENTS

Financial support from CICyT (MAT88-556 and PB89-0642) and Consejeria de Cultura y Bienestar Social (Junta de Castilla y León) is greatly acknowledged. I.M. acknowledges a grant from DGICyT. We thank to the staff of the S.R.S. (Daresbury Lab, U.K.) for their assistance during EXAFS measurements.

REFERENCES

1 T. Ono, Y. Nakagawa, H. Miyata, Y. Kubokawa, Bull. Chem. Soc. Jpn., 1984, 57, 1025.
2 D. Vanhove, S.R.Op, A. Fernandez, M. Blanchard, J. Catal., 1984, 57, 253.
3 Y.C. Lium, G.L. Griffin, S.S. Chan, I.E. Wachs, J. Catal. 1985, 94, 108.
4 C.-L. O'Young, J. Phys. Chem., 1989, 93, 2016.
5 M. Kantschewa, F. Delannay, H. Jeziorowski, E. Delgado, S. Eder, G. Ertl, H. Knözinger. J. Catal. 1984, 87, 482.
6 C.T.J. Mensch, J.A.R. van Veen, B. van Wingerden, M.P. van Dijk, J. Phys. Chem., 1988, 92, 4961.
7 G. Kisfaludi, J. Leyrer, H. Knözinger, R. Prins, J. Catal., 1991, 130, 192.
8 K. Matsumoto, A. Kobayashi, Y. Sasaki, Bull. Chem. Soc. Jpn., 1975, 48, 1009.
9 L. Kihlborg, Arkiv foer Kemi, 1963, 21, 357.

X-Ray Photoelectron Spectroscopic Determination of the Surface Oxidation State of Ceria in UHV

P. G. Harrison[1], D. A. Creaser[1], B. A. Wolfindale[2], and M. A. Morris[2],

[1] DEPARTMENT OF CHEMISTRY, UNIVERSITY OF NOTTINGHAM, UNIVERSITY PARK, NOTTINGHAM NG7 2RD, UK
[2] ICI CHEMICALS & POLYMERS LTD, KATALCO R & T GROUP, PO BOX 1, BILLINGHAM, CLEVELAND TS23 1LB, UK

ABSTRACT

Complex 3d spectra from ceria (CeO_2) have been collected. These spectra display not only an expected spin-orbit doublet but also exhibit pairs of photoelectron features which are thought to be due to electronic transitions which occur during photoemission. By mounting the sample so that it can be heated *in situ* both in vacuum and in a pressure of reducing (H_2 or CO) and oxidising (O_2) gas, photoelectron features due to Ce^{IV} and Ce^{II} could be assessed. Briefly, X-ray photoelectron spectroscopic studies of ceria show a mixture of Ce^{IV} and Ce^{II} at the surface of the analyte. The Ce^{IV} spin orbit coupled features observed are associated with satellite features due to final state effects and the binding energy differences between the primary and shifted peaks agree well with literature values.

INTRODUCTION

Cerium oxide (ceria) has become a rather important material in catalysis[1]. It is used in the cracking of heavy oil over zeolites[2], as an oxygen storage compound in car exhaust systems[3], and as a possible catalyst in the oxidative coupling of methane[4]. As a consequence of its very high melting point, it has considerable potential as both a support and as an active catalyst (in high temperature applications particularly). The chemical activity of ceria arises from the well known defect properties of this material[5]. Ceria can be anion deficient (ie loss of oxygen), and the vacancy electrons at anion vacancies form a small polaron associated with a tetravalent cerium ion. Hence some cerium ion are effectively reduced to the +3 state, and it is this site which is considered to be active in catalysis[4].

To further our understanding of the surface chemistry of ceria, XPS has been used to monitor the relative amounts of +3 and +4 valent cerium ions at the surface as a function of heat and gas treatments. Cerium 3d core level spectra are complex due to fine structure within the spectral envelope[6]. Although many theoretical and experimental studies have been made, absolute assignment of the features has not been made and controversy still exists. In the work reported here, peaks have been assigned by following oxidation and reduction of the surface *in situ*.

EXPERIMENTAL

CeO_2 was prepared by bicarbonate precipitation from aqueous nitrate solutions. The resulting complex oxy-carbonate-nitrate species was dried at 120°C and then aged for 8 hours to effect calcination. Subsequent aging at 750°C for 8 hours effected complete oxidation. The resulting ceria was crushed into a powder and pressed (1 ton pressure) into a high transparency stainless steel gauze. This was mounted onto a high precision manipulator by spotwelding the gauze to copper feedthroughs. The sample was heated by current passing through the copper feedthroughs and the temperature measured using a chromel-alumel thermocouple in intimate contact with the gauge. Within the UHV chamber, a high pressure cell could be wound over the sample for high pressure gas treatments. X-Ray photoelectron (XPS) analysis was carried out using Al-Kα irradiation (180 watts) and spectra were recorded at 50 eV pass ebergy. Although the sample was earthed via the support, charging occurred and binding energies (BE's) are referenced to an adventitious C 1s signal at 285.0 eV when necessary. Much of the work is reported as unreferenced to show that the extent of charging in the sample differs on treatment. For this reason, absolute peak positions cannot be compared to other work. However, the spectra observed are very similar to previously reported data and absolute binding energy measurements are not needed.

RESULTS AND DISCUSSION

Typical results are shown in Figure 1 where a series of Ce 3d photoelectron spectra are shown as a function of treatment. Spectra labelled (a) to (d) have been heated at 300°C, 350°C, 450°C and 550°C for 2 minutes. Spectra labelled (e) and (f) have been heated in 1 atm. (Gauge pressure) of oxygen for 5 minutes at 450°C and 550°C for 5 minutes, respectively. Spectrum (g) has been heated in 1 atm. of hydrogen at 480°C for 5 minutes, spectrum (h) has been heated in 1 atm. of CO at 580°C for 15 minutes. The spectra have not been charge corrected; later the apparent shift in one of the peaks will be used as a measure of the oxidation state at the surface. Finally, spectrum (i) has been heated in oxygen at 1 atm. for 15 hours at 250°C.

The spectrum in Figure 1(a) is typical of spectra reported in the literature for CeO_2. Following convention[6], we have labelled two sets of multiplets u and v ($3d_{3/2}$ and $3d_{5/2}$), respectively. u^{III} and v^{III} are widely assigned to the $3d^9 4f^0$ (5+) photoemission final state, whilst v and v^{II} (u and u^{II}) are usually assigned to a mixing of the $3d^9 4f^2$ and $3d^9 4f^1$ final states. The v/v^{II} and u/u^{II} states exist because of ligand (O 2p) to metal (Ce 4f) charge transfer, producing states of nominal metal charge +3 and +4. The v^I and u^I lines indicated are suspected to derive from cerium ions in a ground state +3 and have been seen to increase on reduction in hydrogen[7].

In Figure 1(a-d) the samples are progressively heated to higher temperatures *in vacuo*. It can be seen that the heating enhances features at binding energies corresponding to v^I and u^I. These assignments were made by measuring positions relative to the u^{III} position. It can also be seen that the broader features v^{II} and u^{II} also increase during this heating cycle.

The features v^{II} and u^{II} are clearly associated with the presence of Ce^{3+}. In Figure 2 spectra recorded from three standards are displayed. Also shown is a spectrum taken from an oxidised cerium metal foil which has been annealed in oxygen (200°C for 10 minutes, 1 atm.), argon bombarded (8KV, 10mA emission, 1 hour) and heated in CO (360°C for 14 hours, 0.6 atm.) to produce a Ce_2O_3 standard. The features indicated as t and s and t* and s* are associated very clearly with Ce^{3+}. Note that because of X-ray inducing charging the t and s and t* and s* features are not quite resolvable at this pass energy.

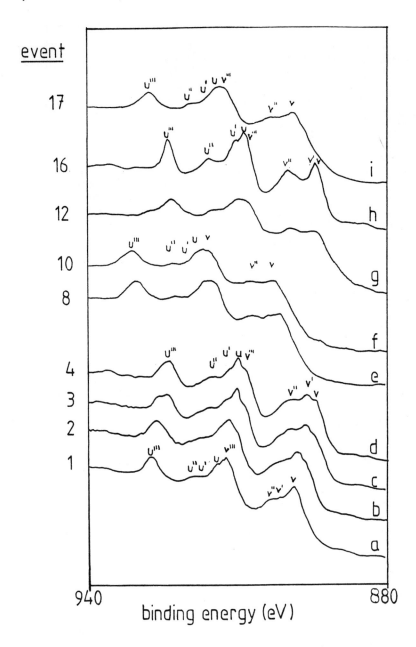

Figure 1. Cerium 3d core level spectra through heating, reduction and oxidation cycles. Full description of the events are described in the text. Various peaks are indicated using nomenclature described in the text.

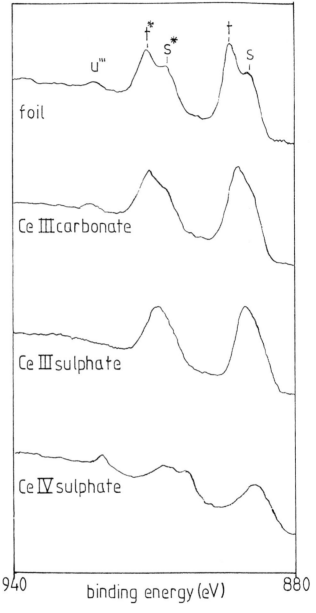

Figure 2. Cerium 3d core level spectra from various standards, as indicated in the diagram. The foil spectrum is from an oxidised foil argon ion bombarded for extended periods at elevated temperatures to bring about reduction. Spectra are charge referenced to an adventitious signal at 285.0eV.

The small amount of Ce^{4+} observed in the spectrum is indicated by u_{3+}^{III}. Measurement of the energy shifts between u_{3+}^{III} and t and between u_{3+}^{III} and s confirm that the peaks indicated as v^l and u^l in ceria spectra are due to the presence of 3+ ions (since the u^{III} to v^l and u^{III} u^l energy differences have identical values). It is also very important to realise that the shifts between u_{3+}^{III} and s^* and between u_{3+}^{III} s are identical to the differences between u^{III} and u and between u^{III} and v. The correspondence of these features in three and four valent cerium ion spectra strongly supports the theory that the features v and u^{II} are shakedown features and arise from O 2p $\longrightarrow$ Ce 4f transitions. The agreement in binding energies suggest that the u and v features are equivalent to the s^* and t features and derive from the same final state. From theoretical work this can be considered as the $3d^9$ $4f^2$ state, although theory also suggests mixing such that this state is more accurately written $3d^9$ $4f^{1.84\ 8}$.

The mixed state corresponding to the $3d^9$ $4f^1$ final state, t and s, (more accurately $3d^9$ $4f^{1.42\ 8}$) is not apparent in the spectrum shown in Figure 1(a) and is only revealed on higher temperature treatments (as u^l and v^l). Instead the other 3d4f mixed state is seen in the Ce^{4+} spectra at a lower binding energy and is indicated as u^{II} and v^{II}. The shift in this peak is not surprising since the peak separation is dependent on the hydridisation of the valence band state and the 4f state, and this will be strongly dependent on the compound structure, etc.

It is important to note that the spectral changes observed on heating provide strong evidence for reduction during this thermal treatment.

The spectra shown in Figure 1(e), (f) and (i) reveal the effects of oxidation. On oxygen treatment distinct features at u^l and v^l are no longer visible as the 3+ component at the surface becomes oxidised. Also obvious is that the u^{II} and v^{II} features (ca. $3d^9$ $4f^1$) are less apparent and almost unresolvable after oxidation. This is also seen in the standard spectrum of cerium(IV) sulphate. It is suggested that these features are more intense in spectra of the surface with the cerium ions in three and four valent states (a mixture of Ce_2O_3 and CeO_2). If the sample is subsequently reduced (after reduction in either H_2 (Figure 1(g)) or CO (Figure 1(h)), the u^{II} and v^{II} features become more apparent as the Ce_2O_3 at the surface increases.

In Figure 3 we show an assignment of photoelectron features in the 3d spectra of Ce_2O_3. Features indicated as (a) are the ground state features due to ionisation of the CeO_2 to a 5+ state during photoemission. Features indicated as (c) are due to a mixed $3d^9$ 4f state corresponding principally to a $3d^9$ $4f^2$ final state. Features marked (b) are the other $3d^9$ 4f mixed final state and approximate to $3d^9$ $4f^1$, deriving from the presence of Ce^{3+} ions in the material. Features due to the Ce^{3+} ions at binding energies between peaks (b) and (c) cannot be resolved in this spectrum. This assignment is definitive evidence of the applicability of the model commonly used to describe these spectra[6-8], but does not support the theroretical work of Thornton and Dempsey[9]. Note that the usability of features indicated (b) on heating to 300°C *in vacuo* (and even on freshly introduced samples) indicates that the surface is reduced either on exposure to UHV or under X-ray irradiation. The relative amounts of 3+ and 4+ components at the surface can be estimated very simply.

Theoretically, it is expected that the increase in abundance of cerium 3+ ions in the sample increases the electrical conductivity[10]. This is due to enhanced motion in bands associated with the polaron states. This effect can be seen quite dramatically in the unreferenced spectra shown in Figure

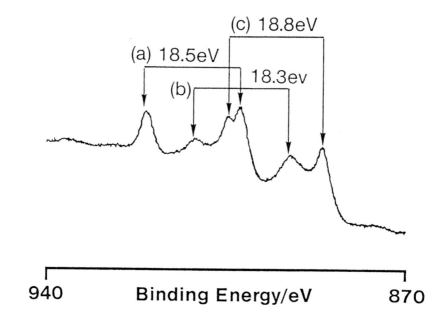

Figure 3. A typical Ce 3d core level spectrum. The spin-orbit coupling energy between peaks is shown. An assignment of each pair is indicated in the text.

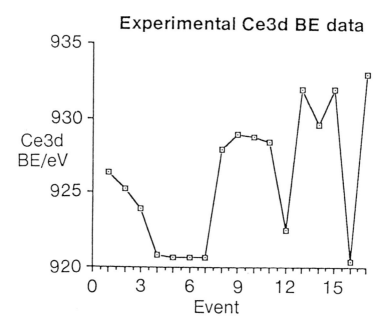

Figure 4. The binding energy of the u^{III} feature through various treatments (see text) plotted against event number.

1(a-i). On reduction (by heating or chemically), the peaks move to lower binding energies, *ie* conduction increases, charging decreases. On oxidation, the sample becomes more insulating (CeO_2 only) and the samples charge more and binding energies are apparently higher. This is also seen as a broadening in the peaks. Data are represented in Figure 4. The x-axis is a scale based on the treatment of the sample (several points are data shown in Figure 1 and are indicated in this diagram). Events 1-7 are thermal treatments, events 8-11, 13, 15, and 17 oxidative treatments, and events 12 and 14 are hydrogen reduction. Event 16 is a CO reduction. The y-axis is a binding energy value from the u^{III} photoelectron work. The position of the u^{III} feature is clearly indicative of the oxidation state. Further confirmation of this can be seen in Figure 5 where the O1s/Ce3d (total peak area ratio is plotted as a function of event. The plots are as expected very similar.

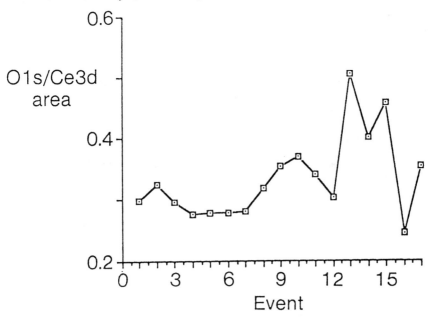

Figure 5. A similar plot to Figure 4 in which the O1s/Ce3d peak area ratio is plotted against event number.

CONCLUSIONS
XPS spectra of CeO_2 have been collected as the sample has been reduced and oxidised "*in situ*". Detailed analysis and comparison with reference spectra from various compounds show that features in the CeO_2 spectra at lower binding energies derive from "shake-off" from the $3d^9\,4f^0$ core hole state. These correspond approximately to $3d^9\,4f^2$ (the extra f electrons arising from O2p levels) states. The observed $3d^9\,4f^2$ feature is a mixture of $3d^9\,4f^2$ and $3d^9\,4f^1$ final states. The observed other 4f final state (*ca* $3d^9\,4f^1$) is at an apparent binding energy different to that observed in Ce_2O_3 spectra. The Ce_2O_3/CeO_2 ratio observed at the surface of the sample strongly affects the absolute binding energy positions of the Ce 3d features. The increase in conductivity as the sample is reduced (Ce_2O_3 enhanced) is reflected in lower charging of the sample and a concomitant reduction in binding positions and sharpening of the peaks.

ACKNOWLEDGEMENTS
One of us (D.A.C.) would like to thank the SERC and ICI plc for a CASE studentship. The authors thank M. Fowles, K.C. Waugh and W.C. Mackrodt for stimulating discussions. The ICI Strategic Research Funding Scheme is acknowledged for project support.

REFERENCES
1. R.J.H. Voorhover, "Advanced Materials in Catalysis", Academic Press New York, 1973, p 129.
2. T.M. Tri, J. Massardier, P. Gallezot and B. Inelik, Proceedings of the 7th International Congress on Catalysis. Tokyo, T. Seiyama and K. Tanabe (eds), Elsevier, Amsterdam, 1980.
3. K.C. Taylor, "Catalysis: Science and Technology", J.R. Anderson and M. Boudart (eds), Springer Verlag, Berlin, 1984.
4. V.R. Choudhary and V.H. Rane, J. Cat., 1991, **130**, 411.
5. A.G. Ashirs, Chem. Letters, 1990, **6**, 67.
6. P. Burroughs, A. Hamnett, A.F. Orchard and G. Thornton, J. Chem., Soc., Dalton Trans., 1976, 1686.
7. A. Laachir, V. Perrichon, A. Badri, J. Lamotte, E. Catherine, J.G. Lavalley, J.E.R. Fallah, L. Hilaire, F. le Normand, E. Qoemere, G.N. Sauvion and O. Tovret, J. Chem. Soc., Faraday Trans. I, 1991, **87**, 1601.
8. T. Nakano, A. Kotani and J.C. Parlebas, J. Phys. Soc. Japan, 1987, **56**, 2201.
9. G. Thornton and M.J. Dempsey, Chem. Phys. Lett., 1981, **77**, 409.
10. B.S. Chiou, J. Am. Ceram. Soc., 1990, **28**, 866.

Characterisation and Testing of Carbon Monoxide Oxidation Catalysts

N. Jorgensen
CATALYST UNIT, MATERIALS CHEMISTRY DEPARTMENT,
AEA TECHNOLOGY, BUILDING 429, HARWELL LABORATORY,
OXFORDSHIRE OX11 0RA, UK

ABSTRACT

The preparation conditions for precious metal/tin oxide catalysts were optimised for maximum carbon monoxide/oxygen reactivity. This was achieved by controlling the tin digestion, the peptisation to form a sol, the calcination process and the method of adding the precious metals. In addition, extensive studies of the tin oxide structure were carried out over the temperature range 20 to 500°C in air, carbon monoxide or hydrogen environments using Raman spectroscopy and X-ray diffraction.

1. INTRODUCTION

Carbon monoxide oxidation catalysts, active at room temperature, can be used in several applications, including escape hoods[1], miner's self rescuers[2], high powered carbon dioxide lasers[3], air conditioning systems[4] and cigarette filters[5].

The applications of closed cycle carbon dioxide lasers range from highly technical observation systems to heavy duty steel cuting tools. During operation the carbon dioxide can dissociate to such an extent that the operation of the laser is impaired and halted. If a suitable catalyst is incorporated into the stream of circulating gas, the carbon monoxide can be recombined with oxygen forming carbon dioxide.

For air conditioning[4] or escape systems, the requirement is to remove, by reaction, carbon monoxide from the breathed air. Hoods or masks for use by people escaping from fires in aircraft, hotels or coal mines could incorporate a carbon monoxide oxidation catalyst[6]. If the carbon monoxide is thus removed from the breathed air, the likelihood of survival of the person is much greater than would otherwise be the case. Coal miners in the UK carry a self-rescuer when underground. Currently the carbon monoxide removal component of these masks is hopcalite. UK Government bodies have carried out trials using hoods intended for escape from fires in aircraft, but have so far concluded that the systems tested do not meet their requirements.

Because of the range of possible uses of carbon monoxide oxidation catalysts which can operate at temperatures about 20°C, there is considerable interest in the structure, stability and performance of such catalysts[7-11]. The requirements are for a material which has good carbon monoxide oxidation performance, maintains this performance over a long time period, can tolerate the presence of other species in the gas stream and must be stable during transportation and non-working (eg. storage) periods.

High surface area precious metal/tin oxide catalysts display good activity towards the carbon monoxide oxidation reaction[2,3]; therefore the aim of this investigation was to determine the effect of catalyst preparation conditions and catalyst structure on the stability and performance of such materials.

2 EXPERIMENTAL

Catalyst Preparation

Several preparation methods have been used for tin oxide supports[12-16]. Metastannic acid can be obtained commercially or prepared by controlled reaction of tin metal with nitric acid. In this work it was desirable to control the particle size by varying the concentration of the acid used.

A tin oxide sol was prepared by peptising the metastannic acid using an organic base. The metastannic acid was washed until the supernatant liquid had a conductivity of less than 4.5 mS prior to peptisation. The resulting sol had a typical oxide content of 300g l^{-1}, with an average particle size of about 200nm (measured using a Malvern Zetasizer). The sol was then dried and the oxide powder obtained by calcining at 350°C.

The precious metals were added as a salt solution to either the sol or the calcined tin oxide powder. Platinum and palladium were added such that after calcination at 350°C their contents were in the range 0 to 3% of tin oxide weight.

Activity Testing

Samples of powdered catalyst were placed in glass reactor tubes and reduced in hydrogen/helium. Each catalyst was then stabilised in oxygen. Activity testing was carried out in flowing gas mixture consisting of 0.5% oxygen, 1.0% carbon monoxide and 98.5% nitrogen at a space velocity of 36,000h^{-1}. The CO and O_2 concentrations in the exhaust stream were constantly monitored in order to determine catalyst efficiency with time.

Raman Spectroscopy

These experiments were carried out using a Spex Triplematic monochromator fitted with an intensified 1024 photo-diode detector. The laser beam sources were capable of being adjusted over a wavelength range 441 nm (He/Cd) to 514 nm (Ar ion). The power of the source beam was 100 mW but was reduced to 10 mW at the sample. All of the high temperature, controlled atmosphere, experiments were carried out using a Stanton Redcroft HSM 5 hot stage.

X-Ray Diffraction

The room temperature X-ray diffraction experiments were carried out using a Siemens fully automated D500 X-ray diffractometer with copper K_α radiation and analysed using a secondary monochromator through a scintillation counter. The high temperature experiments were carried out using an AEA

Technology precision apex goniometer with a GTP Engineering high temperature attachment. The diffracted beam passed through a curved graphite crystal secondary monochromator.

3 RESULTS AND DISCUSSION

Tin Oxide Stability

Displayed in Figure 1 is the TGA plot for the gel derived from the base peptised tin oxide sol. It is apparent from this that the material should be calcined at 350°C, otherwise residues of the base remain. Calcining at higher temperatures resulted in greater crystallinity but a loss of surface area.

A sample of un-peptised tin oxide was enclosed in a Raman cell and heated in 100°C steps, each for 30 min, from room temperature to 500°C. It can be seen from Figure 2 that the A_{1g} band moved from 635 cm^{-1} at 20°C to 626 cm^{-1} at 500°C, and that the B_{2g} band similarly moved from 776 cm^{-1}. Similar, but smaller, band shifts were observed when heating in hydrogen rather than air.

These spectra were similar in appearance to those for single crystals[17] and textured material[18] and the wavenumber shifts to those for single crystals[19].

Activity Measurements

Initial experiments indicated that catalysts prepared by the incipient wetness technique were more active than for the mixed salt solution and sol method with the same amount of precious metal added. Experiments were not carried out with the precious metal content maximised; therefore this observation is probably due to the greater concentration on the surface resulting from the incipient wetness technique. All experiments described below used catalyst prepared by the incipient wetness technique.

Catalysts containing 0, 1, 2 or 3% platinum and palladium were prepared and tested at room temperature. The activity measurements suggest that the samples fall into three categories. The first group contained the 2% Pd, 2% Pt, 1% Pt/2% Pd samples (Figure 3). Each of these activated, especially the 2% Pt sample, but then deactivated later in the test. The second group containing the 3% Pt/1% Pd, 2% Pt/1% Pd and 1% Pt/3% Pd samples sustained activity at a reasonable level for the duration of the test. The third group containing the 2% Pt/2% Pd, 2% Pt/3% Pd, 3% Pt/2% Pd and 3% Pt/3% Pd samples continued activating during the test (Figure 4).

Catalytic activity was increased by adding more Pt or Pd, but was more critically dependent on the Pt content. In many cases the catalysts were operating in excess of 90% conversion; therefore the apparent effect of increasing precious metal loading was to sustain the high activity for a longer period of time. This was also observed for powdered or coated monolith catalysts tested at 60°C, where activity was sustained for longer times. At high gas flowrates, the extent of deactivation was more dependent on the workrate than on the total quantity of CO oxidised.

Deactivated catalysts can normally be regenerated by re-reducing, or even by flushing with an inert gas, therefore it is thought that deactivation occurs due to a self-poisoning type mechanism rather than by structural changes in the catalyst surface.

272   Catalysis and Surface Characterisation

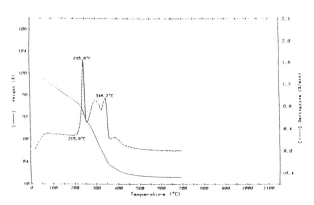

Figure 1 TGA plot for calcination of tin oxide

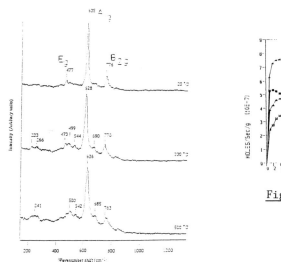

Figure 2 Raman spectra of tin oxide in air

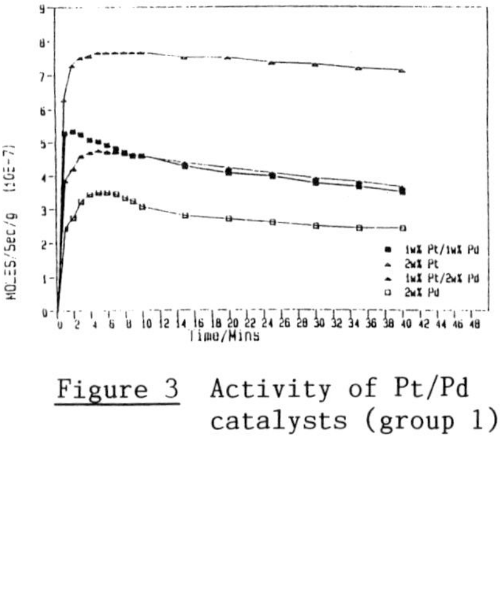

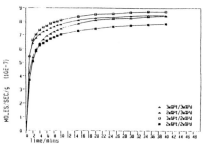

Figure 3 Activity of Pt/Pd catalysts (group 1)

Figure 4 Activity of Pt/Pd catalysts (group 3)

Poison Resistance

For catalysts oxidising carbon monoxide at 20°C, water present in modest quantities acted as an activity promoter, perhaps by hydroxylating the tin oxide surface or by suppressing self-poisoning. The hopcalite used presently in miner's self-rescuers is deactivated in the presence of water[2].

Acetone acted as a catalyst poison, with activity only restored after re-reduction, while an injection of hexane temporarily decreased activity and 1,1,1-trichlorethane acted as a permanent poison. The effect of such poisons was much smaller if the catalyst was operated at higher temperatures. For operation at 20°C, either part of the catalyst bed should be sacrificed as a filter or guard beds, carbon filters etc. should be used if such poisons are likely to be present in the gas stream.

Adsorbed Species

Experiments were conducted using a high temperature Raman cell in order to determine the influence of carbon monoxide (5% in helium) on the tin oxide and the catalysts. With tin oxide alone a band appeared at about 1600 cm^{-1} just below 200°C. On heating further towards 500°C the band intensified (Figure 5), and persisted on cooling to 20°C. Such bands have been attributed to the CO stretch in, for example, a carbonate adsorbed on silica or alumina in similar environments.

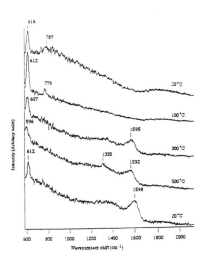

Figure 5 Raman spectra of tin oxide in CO.

For a precious metal/tin oxide catalyst heated in CO/helium a similar band appeared at 100°C, a considerably lower temperature than for the tin oxide alone. As for the tin oxide, this band grew in intensity with increasing temperature up to 500°C. However, the band disappeared on cooling to 20°C, probably due to desorption or catalytic decomposition under the influence of the precious metal.

This influence of precious metals or transition metals on the presence of adsorbed species on support oxide surfaces has been observed previously[20-22], for example in the formation of NCO from CO and NO on a metal, followed by migration onto the support. The catalytic reactions forming the NCO could occur directly on the oxide surface, but at a significantly higher temperature.

4 CONCLUSIONS

To achieve optimum catalyst performance the preparation of the tin oxide sol must be carefully controlled. If the tin oxide is calcined at 350°C a stable, high surface area, support is obtained. Greater crystallinity can be induced by heating to higher temperatures, but this is accompanied by a decrease in surface area. It was noted that a 2% Pt catalyst had an activity several times greater than for a 2% Pd catalyst, and that the higher Pt/Pd loaded catalysts had extended lifetimes, expecially at greater workrates.

Species displaying a CO stretch at about 1600 cm^{-1} were formed on tin oxide above 200°C and on catalyst containing precious metal above about 100°C. The band persisted on cooling to 20°C for the tin oxide, but disappeared for the catalyst. This was attributed to the catalytic effect of the precious metal.

Catalysts displaying the level of activity required at 20°C to maintain the carbon monoxide and oxygen levels below critical values in an operating high powered carbon dioxide laser or the CO level in an escape system could be obtained by carefully controlling the materials preparation and catalyst pretreatment conditions.

ACKNOWLEDGEMENTS

I would like to thank Mr C F Sampson for his work, Dr C Johnson for the Raman data, Mr F Cullen for the X-ray diffraction studies, and the UK Ministry of Defence, RSRE Malvern, for supporting part of this work.

REFERENCES

1. C F Sampson and N J Gudde, NASA Conference Publication 2456, 1987, 65.
2. C J Wright, B C Tofield and C F Sampson, UK Patent GB 2 141 349 B, 1987.
3. C F Sampson and N Jorgensen, NASA Conference Publication 3076, 1990, 57.
4. R Yaparpalvi and K T Chuang, Ind.Eng.Chem.Res., 1991, 30, 2219.
5. M D Shannon, R L Lehman, J L Resce, O P Surin, J T Mears, D M Riggs and E G Farrier, International Patent Application Number PCT/US89/01080, 1990.
6. House of Commons Transport Committee, Session 1990-91, 'Aircraft Cabin Safety, Volume I, Volume II', 1990.
7. NASA Conference Publication 2456, 1987.
8. NASA Conference Publication 3076, 1990.
9. T P Moser, European Patent Application 0 421 169 A1, 1991.

10. S D Gardner, G B Hoflund, B T Upchurch, D R Schryer, E J Kielin and J Schryer, J.Catalysis, 1991, 129, 114.
11. P A Sermon, V A Self and E P S Barrett, J.Chem.Soc., Chem.Comm., 1990, 1572.
12. E S Lane, UK Patent GB 2 125 3923, 1986.
13. E S Lane, UK Patent Application GB 2 126 205 A, 1984.
14. C J Wright and C F Sampson, UK Patent GB 2 134 004 B, 1986.
15. E S Lane, UK Patent GB 2 134 413 B, 1986.
16. D L Segal, S R Daish and J L Woodhead, UK Patent Application GB 2 155 915 A, 1985.
17. R S Katigar, P Dawson, M I Hargreave and G R Wilkinson, Phys.C.Solid Phys., 1971, 4.
18. J Geurts, S Rau, W Richter and F J Schmitte, Thin Solid Films, 1984, 121, 217.
19. P S Peercy and B Morosi, Phys.Rev.B, 1973, 7(6).
20. F Solymosi, L Volgyesi and J Rasko, Z.Phys.Chem.Neue Folge, 1980, 120, 79.
21. C Johnston, N Jorgensen and C H Rochester, J.Chem.Soc. Faraday Trans.1, 1989, 85(5) 1111.
22. C Johnston, N Jorgensen and C H Rochester, J.Chem.Soc. Faraday Trans.1, 1988, 84(6), 2001.

Oxidation Studies on Pt–Sn Supported Catalysts

J. W. Curley and O. E. Finlayson
SCHOOL OF CHEMICAL SCIENCES, DUBLIN CITY UNIVERSITY, DUBLIN 9, IRELAND

A series of Pt and Pt - Sn catalysts with nominally 5 wt% Pt and varying Sn contents (0 - 3 wt%) supported on Al_2O_3 were characterised by H_2 chemisorption and X-ray diffraction measurements. SnO_2 supported Pt (1 and 5 wt% Pt) were also prepared. The activity of these catalysts for butane oxidation was determined. On Al_2O_3 supported catalysts, the addition of $\leq$ 1 wt% Sn significantly increased the uptake of H_2, while with further additions to 3 wt% Sn, the H_2 uptake decreased. The Pt/SnO_2 catalysts did not adsorb detectable quantities of H_2. The order of increasing light-off-temperature for the series of catalysts for oxidation of butane was as follows:

Pt/SnO_2 < Pt/Al_2O_3 < $Pt-Sn/Al_2O_3$.

On aging to 1073 K, the light-off-temperature for butane oxidation on Pt-Sn/Al_2O_3 catalysts remained unchanged while that of Pt/Al_2O_3 and Pt/SnO_2 increased by 45 K and 71 K respectively.

1 INTRODUCTION

Platinum supported on tin oxide is known to catalyse a number of oxidation reactions including oxidation of methanol and carbon monoxide.[1] Bimetallic Pt-Sn catalysts are important in the hydrocarbon reforming industry, due to their improved resistance to deactivation by coke formation and increased selectivity towards aromatic products.[2,3]

The effect of Sn addition on the activity of Pt/Al_2O_3 for total oxidation of i-butane is examined in this study and compared to that of Pt/SnO_2; also the effect of thermal aging on the activity of the samples was examined.

2 EXPERIMENTAL

Catalyst Preparation

A series of Pt-Sn/Al_2O_3 catalysts, with nominally 5 wt% Pt and varying Sn contents (0 - 3 wt%) were prepared by co-impregnation with aqueous solutions of $H_2PtCl_6 \cdot 6H_2O$ and $SnCl_4 \cdot 5H_2O$. The Al_2O_3 fibre mat "Saffil" was supplied by ICI (Runcorn). The following samples were prepared and dried at 313 K for 16 hr: 5%Pt/Al_2O_3 (PA), 5%Pt - 3%Sn/Al_2O_3 (PS3), 5%Pt -1%Sn/Al_2O_3 (PS1), 5%Pt - 0.3%Sn/Al_2O_3 (PS0.3). Two 5%Pt-Sn/Al_2O_3 samples were prepared on pretreated Al_2O_3 with either 1 or 0.3 wt%Sn (PS1(t), PS0.3(t) respectively). The Al_2O_3 was prewashed with 0.1 M HNO_3 followed by H_2O until the washings were neutral, and dried at 343 K for 16 hr prior to impregnation.

Pt/SnO_2 catalysts, with either 5 wt% (PSO) or 1 wt% (P1SO) Pt, were prepared by addition of aqueous solutions of $H_2PtCl_6 \cdot 6H_2O$ to SnO_2 powder, followed by drying at 323 K for 16 hr. One sample of SnO_2 was pretreated in 10 M NaOH for 0.5 hr at 363 K before impregnation with 5 wt% Pt (PSO(t)) to increase the number of OH sites on the surface and hence the take up of Pt.[4]

All samples were calcined at 773 K for 2 hr except PA which was calcined at 903 K for 0.25 hr. Aging was at 1073 K for 8 hr in static air. The loadings of Pt and Sn on the catalysts were confirmed by atomic absorption spectroscopy to be as the nominal values.

Catalyst Characterisation

H_2 uptake was determined by H_2 pulse chemisorption using a Micromeritics Pulse Chemisorb 2700 with purified gases (Ar through a BOC RPG4, and H_2 through a Johnson Matthey EP1 purifier). Sample PA was reduced in H_2 at 523 K for 1 hr followed by 623 K for 2 hr and samples PS and PSO at 573 K for 0.33 hr. Chemisorption values (±4%) were determined at 308 K. The limit of detection of H_2 uptake was 0.005 µmol $H_2 g^{-1}$. XRD measurements were carried out using a Philips 1400 series X-ray diffractometer with a Cu Kα source.

Activity Measurements

The activity of the catalysts was determined by measuring the light-off-temperature (LOT) for i-butane oxidation using a differential scanning calorimeter (DSC). The LOT is defined in this work as the minimum temperature at which an exothermic process occurred. An air / i-butane mixture (ratio 32:1) was passed over 1 mg of sample, placed in the DSC furnace, at a rate of 19.8 $cm^3 min^{-1}$. After achieving a homogenous gas mixture in the DSC furnace (after approx 0.18 hr), the furnace was heated to 573 K at a rate of 10 Kmin^{-1} and held at that temperature for 0.5 hr prior to cooling to room temperature. LOT was measured three times for each sample; the first LOT was approx 20 K higher than that for runs 2 and 3, which agreed within ± 2 K. The average values of runs 2 and 3 are used to determine LOT.

3 RESULTS AND DISCUSSION

The results for H_2 uptake are given in Table 1. A Pt:H adsorption stoichiometry of 1:1 is assumed in all calculations, even though there is some doubt if this is the case for Pt-Sn systems.[5] The H_2 uptake increased for addition of ≤1 wt% Sn relative to PA and decreased for 3 wt% additions. PSO samples were found not to chemisorb detectable amounts of H_2. PS1(t) had a very high H_2 uptake, 57% greater than PS1. However, with PS0.3(t), the H_2 uptake was 48% lower than for the corresponding untreated support. PA prepared using pretreated alumina gave well dispersed metal particles, with H_2 uptake of 68 $\mu mol H_2 g^{-1}$.

After aging at 1073 K for 8 hr, a decrease of approx 90% in H_2 uptake was observed in PA, PS1, PS0.3 and PS0.3(t), (Table 1). Pretreatment of the support did not effect the decrease in H_2 uptake.

The LOT results for butane oxidation are given in Table 1 for unaged and aged samples. Of the unaged samples, PSO and PA had the lowest LOT at 428 K and 433 K respectively. Decreasing the Pt content to 1 wt% (P1SO) resulted in an increase in LOT relative to PSO. Of the Pt-Sn/ Al_2O_3 catalysts the sample PS3 gave a lower LOT than either PS1 or PS0.3. The LOT was reduced by approx 10 K for PS1(t) and PS0.3(t). PSO(t) gave a LOT approx 15 K greater than PSO. The order of increasing LOT was as follows:

PSO < PA < PSO(t) << PS3 < PS1(t) = PS0.3(t) < P1SO = PS0.3 ≈ PS1

On aging, the LOT for PA and PSO increased by 45 K and 71 K respectively. Aged PS samples (PS3, PS1, PS0.3, PS1(t), PS0.3(t)) gave relatively unchanged LOT. The LOT of all the aged samples was in the range 483 K ± 16 K with the exception of P1SO at 513 K.

X-ray diffraction analysis was carried out on the Al_2O_3 support material, PA, PS and PSO (Fig 1). PA (Fig 1b) shows peaks due to the support (Fig 1a) and metallic Pt. The (311) peak for Pt at 2θ = 81.3° is the only Pt diffraction line clearly distinguishable from the support. A similar pattern was obtained for an aged PA, except that the peaks due to metallic Pt increased in intensity. For sample PS3 (Fig 1c), no peaks due to Pt or Sn could be distinguished from the background of the support

Table 1 Chemisorption and LOT Results

Sample	H$_2$ Uptake /μmol H$_2$ g^{-1}		LOT /K	
	Unaged	Aged	Unaged	Aged
PA	42	2	433	478
PS3	24	nd	468	468
PS1	58	2	488	475
PS0.3	71	10	483	488
PS1(t)	94	nd	478	482
PS0.3(t)	48	3	478	478
PSO	ND	nd	428	499
PSO(t)	ND	nd	443	498
P1SO	ND	nd	483	513

nd = not determined, ND = not detectable

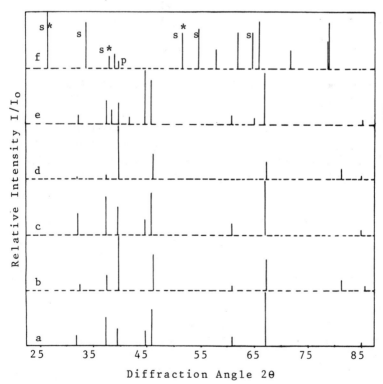

Fig. 1. XRD diffraction pattern: (a) Al$_2$O$_3$ support; (b) PA; (c) PS3; (d) PS3 aged; (e) PS3 reduced at 773 K; (f) PSO (s = SnO$_2$, p = Pt, * = peak ht/5).

peaks even after reduction at 773 K (Fig 1e). However, large quantities of Pt were detected in PS3 after aging (Fig 1d) suggesting that the Pt particles are too small to be detected in unaged PS3. In reduced samples of PS3 and PA (reduced at 773 K), there were three peaks at $2\theta = 38.5°$, $41.7°$ and $65.1°$ which could not be assigned.

Fig 1f shows XRD of PSO. The peaks due to SnO_2 and metallic Pt are labelled; the remainder of the peaks did not correspond to metallic Sn, the alloys PtSn, $PtSn_4$, $PtSn_2$, Pt_2Sn_3, or Pt_3Sn or to Pt or Sn oxides.

The chemisorption results on PS samples in this work agree with those of Balakrishnan and Schwank[6] in that H_2 uptake decreased with increasing Sn content. Several explanations for this effect have been put forward: alloy formation; loss of Pt surface area due to Sn enrichment of the surface; trapping of Pt by a $Sn-Al_2O_3$ matrix (thereby making the Pt atoms inaccessible for chemisorbing gases); or a geometric effect in which the Sn atoms separate Pt atoms giving fewer adjacent Pt sites onto which H_2 can chemisorb dissociatively.[6] No evidence for the presence of Pt-Sn alloys was found in this work in either reduced or calcined Pt-Sn/Al_2O_3 samples. Srinivasan et al.[7] monitored alloy formation by XRD, after *in situ* reduction, and found evidence for Pt-Sn alloys in Pt-Sn/Al_2O_3 containing 0.6 wt% Pt and 1 wt% Sn.

The ability to take up H_2 did not directly relate to the LOT as the reduced catalyst surfaces examined by H_2 chemisorption probably bears little resemblance to the calcined surfaces involved in the oxidation reactions.

The differences in the activities between unaged PSO, PA and PS may partially be explained from TPR studies. Hughes and McNicol[8] studied Pt/SnO_2 with up to 10 wt% Pt by TPR. They found that after calcination in air at 773 K, the TPR profiles of Pt/SnO_2 were devoid of any Pt features and they concluded that Pt reduction could be occurring at the same temperature as SnO_2, in the region of 680 K, and that the calcination step led to the formation of a compound between Pt and SnO_2 which was difficult to reduce.[8] TPR studies on Pt-Sn/Al_2O_3 systems after calcination at 773 K, showed peaks due to the reduction of Pt and Sn at >553 K, but the temperature for Pt reduction had increased by approx. 15 K compared to Pt/Al_2O_3.[4,9] Therefore it is suggested that Pt in PSO is in a different form to Pt in either PA or PS and that this Pt species on the SnO_2 is more active for butane oxidation and has a reduction temperature greater than 573 K used in chemisorption measurements.

The increase in LOT for PSO after aging may be due to Pt sintering. Thermal degradation of the support material may also be a factor as the BET surface area of PSO decreased by approx 40% from

9.5 m^2g^{-1} to 5.8 m^2g^{-1} after aging. In the case of PA, the increase in LOT on aging is due to Pt sintering as the BET surface area of Al_2O_3 support is unchanged after heating at 1073 K for 24 hr.[10]

For unaged samples PS had higher LOT than PA or PSO. This lower activity may be due to electronic or geometric effects brought about by the presence of Sn. The electronic properties of Pt are changed due to electron withdrawal by Sn(II) on the Al_2O_3 surface[11] giving Pt which is more electron difficient in Pt-Sn/Al_2O_3 than in Pt/Al_2O_3. Sn may also cover the Pt atoms on the surface and hence block active sites.[2,3]

The resistance to aging shown by the PS samples could be due to a geometric effect, in which the Sn decreases the amount of Pt sintering. It is possible that Sn(II), stabilised on the Al_2O_3 surface, in the immediate vicinity of the Pt, acts as physical barriers, preventing the coalescence of Pt particles. Meitzner et al.[12] proposed a model in which the Sn particles are attached to the Al_2O_3 surface and clusters of Pt are anchored to the support through Sn(II).

It is interesting to note that the LOT of unaged and aged PS is relatively unchanged after aging while the LOT of PA and PSO increases to the level of PS.

REFERENCES

1. G.B. Hoflund, 'Preparation of Catalysts III', G. Poncelet, P. Grange and P. Jacobs, (Eds), Elsevier, Amsterdam, 1983 p.91.
2. F.M. Dautzenberg, J.N. Helle, P. Biloen and W.M.H. Sachtler, J.Catal., 1980, 63, 119.
3. J. Volter, G. Lietz, M. Uhlemann and M. Hermann, J.Catal., 1981, 68, 42.
4. M. Watanabe, S. Venkatsen and H.A. Laitinen, J.Electrochem.Soc., 1983, 130, 59.
5. H. Lieske and J. Volter, J.Catal., 1984, 90, 96.
6. K. Balakrishnan and J. Schwank, J.Catal., 1991, 127, 287.
7. R. Srinivasan, R.J. De Angelis and B.H. Davis, J.Catal., 1987, 106, 449.
8. V.B. Hughes and B.D. McNicol, J.Chem.Soc. Faraday I, 1979, 75, 2165.
9. A. Sachdev and J. Schwank, 'Proceedings, 9th International Congress on Catalysis, Calgary, 1988', M.J. Phillips and M. Ternan, (Eds), Chem. Institute of Canada, Ottawa, 1988, p.1275.
10. J.W. Curley, M.J. Dreelan and O.E. Finlayson, Catalysis Today, 1991, 10, 401.
11. R. Burch, J.Catal., 1981, 71, 348.
12. G. Meitzner, G.H. Via, F.W. Lytle, S.C. Fung and J.H. Sinfelt, J.Phys.Chem.,1988, 92, 2925.

Characterization by IR and Raman Spectroscopy of Pt/Alumina Catalysts Containing Ceria

M. S. Brogan[1], J. A. Cairns[2], and T. J. Dines[1]
[1] DEPARTMENT OF CHEMISTRY, THE UNIVERSITY, DUNDEE DD1 4HN, UK
[2] DEPARTMENT OF APPLIED PHYSICS, ELECTRONIC AND MANUFACTURING ENGINEERING, THE UNIVERSITY, DUNDEE DD1 4HN, UK

INTRODUCTION

Ceria is now established as an important component of a modern 3-way exhaust catalyst containing Pt/Rh/Alumina. This is due primarily to its ability to act as an oxygen storage component (1,2,3) and to stabilize γ-alumina (4,5). There is also direct evidence to show that ceria promotes CO oxidation (6,7), while the results of its effect on hydrocarbon oxidation are less clear (8,9,10). The purpose of the present investigation is to examine, by spectroscopy techniques, the interaction between Pt and an alumina support as compared with an alumina/ceria support.

EXPERIMENTAL

Catalyst samples containing 2 wt. % Pt were prepared by impregnation of an $\dot{\eta}$-alumina sol (100 m^2 g^{-1}) with $[Pt(NH_3)_4]Cl_2$ and dried for 18 h at 353 K. Aqueous ceria sol (NO_3:CeO_2 ratio 0.28) was prepared by deaggregation of cerium hydrate using nitric acid as a peptizing agent. Catalysts containing ceria were prepared by adding a measured volume of ceria sol (25 m^2g^{-1}) to the alumina before impregnation of the Pt. The Pt/ceria catalysts were prepared by impregnation of ceria sol with $[Pt(NH_3)_4]^{2+}$.

Self-supporting catalyst discs (100 mg, 2.5 cm diameter) were pressed, at 10^6 N m^{-2}, and mounted in an infrared cell attached to a conventional glass vacuum apparatus. They were initially calcined in flowing oxygen at 723 K for 1 h prior to reduction at 573 K in hydrogen (3.5% in argon), unless otherwise stated. Infrared spectra were recorded on a Perkin Elmer 681 dispersive spectrometer linked to a PE3600 data station. The infrared study of CO adsorption involved measurement of spectra of adsorbed CO at an equilibrium pressure of 13.3 kN m^{-2}, which was sufficient to ensure complete surface coverage of the available Pt sites.

Raman spectra were excited by the 488.0 and 514.4 nm lines of an argon-ion laser (Coherent Radiation Innova 90-6) and recorded on a Spex 1403 spectrometer controlled by a DM1b computer. Catalyst samples in the form of self-supporting discs were illuminated in the 90° configuration. In order to avoid thermal decomposition the laser power was maintained at less than 50 mW at the sample. No changes in the spectra were evident even after continuous irradiation for several hours. Detection was by standard photon counting techniques using a Hamamatsu R928 photomultiplier. Data were acquired at 2 cm^{-1} intervals with 1 s integration time and the spectral slit width was 2-3 cm^{-1}. In each case the spectra were averaged over at least 4 scans to improve the signal-to-noise ratio, and corrected for the spectral sensitivity of the instrument. The wavenumber calibration of the spectrometer was ± 1 cm^{-1}, established by reference to the Ne emission spectrum.

Results and Discussion

Raman spectra for the Pt/alumina catalyst before and after calcination are shown in figures 1(a) and (b) respectively. The ν(Pt-N) symmetric and asymmetric stretching vibrations at 538 and 568 cm^{-1} together with the δ(NPtN) deformation at 320 cm^{-1} are attributable to the [Pt(NH$_3$)$_4$]$^{2+}$ (11). After calcination (figure 1b) these bands are no longer observed, indicating decomposition of the [Pt(NH$_3$)$_4$]$^{2+}$ precursor. Moreover there are no bands which could be attributed to any Pt oxide species. It is postulated that surface Pt oxide species on the alumina support would give rise to relatively weak Raman bands which, under present experimental conditions, are too weak to be observed.

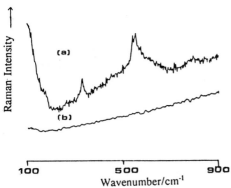

Figure 1 Raman spectra of Pt/alumina (a) before and (b) after calcination in flowing O$_2$ at 723 K for 1 h.

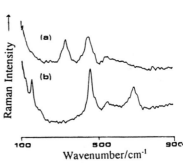

Figure 2 Raman spectra of Pt/alumina with a 20 wt. % ceria loading (a) before and (b) after calcination in flowing O$_2$ at 723 K for 1 h.

Otto et. al.(12) recorded Raman spectra of Pt/alumina catalysts having varying Pt loadings, from 1.2 to 30 wt. %, and observed bands at 125, 335 and 590 cm^{-1} which were attributed to dispersed phase Pt, i.e. an oxide of Pt in a 2 dimensional structure. In order to observe these bands the spectra were recorded with 20 minute integration times and a laser power of no more than 10 mW. Such experimental conditions would be impractical using the equipment employed here. For catalysts with a Pt loading of greater than 5% the absence of Raman bands was taken as evidence for the formation of particulate Pt.

For ceria-containing catalysts the Raman results are quite different. Figure 2a shows the Raman spectrum of a Pt/alumina/ceria catalyst having a ceria loading of 20 wt. %, prior to calcination. In contrast to figure 1a symmetric and asymmetric ν(Pt-N) bands at 552 cm^{-1} are much weaker in intensity than the deformation band at 332 cm^{-1}. This suggests that, either during impregnation of the Pt(NH$_3$)$_4$]$^{2-}$ or during the drying stage, the environment of the Pt is influenced by the presence of ceria. The ceria precursor gives rise to a band at 452 cm^{-1} which after calcination shifts to 462 cm^{-1} as the hydroxylated cerium changes to ceria. This band is assigned to a t$_{2g}$ Raman active mode characteristic of a lattice possessing the fluorite structure (13).

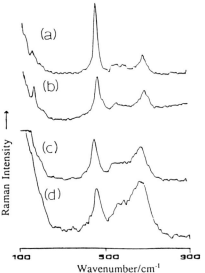

Figure 3 Raman spectra of calcined (a) Pt/ceria, Pt/alumina with a ceria loading of (b) 20 wt. %, (c) 10 wt. % and (d) 5 wt. %

In the spectrum of the calcined Pt/alumina/ceria catalyst two new bands were observed at 560 and 690 cm^{-1}. The spectrum of the calcined Pt/ceria catalyst (figure 3a) is almost identical. Therefore, it can be concluded that the surface species giving rise to these bands is independent of the presence of the alumina. This conclusion is strengthened by the observation of the same bands for calcined Pt/alumina catalysts having ceria loadings of 10 and 5 wt. % (figures 3c and d, respectively). Although the 462 cm^{-1} band, due to ceria, increases in intensity with ceria loading the relative intensity of the other bands remain unaltered.

Figures 4a and b show the spectra of ceria before and after calcination respectively. Following calcination the broad band at 452 cm^{-1} is replaced by a narrower band at 462 cm^{-1}. In contrast, the calcined alumina surface exhibits no Raman active modes. This is consistent with other work (14,15) which suggest that many of the phases of alumina have no Raman active modes. Furthermore, the Raman spectrum of a calcined alumina/ceria sample (figure 5) indicates no evidence of an alumina-ceria interaction, even at a calcination temperature of 873 K.

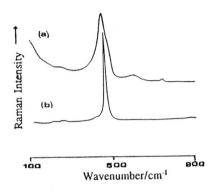

Figure 4 Raman spectra of ceria (a) before and (b) after calcination in flowing O_2 at 723 K for 1 h.

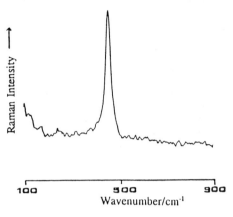

Figure 5 Raman spectrum of alumina/ceria (10 wt. %) calcined in flowing O_2 at 873 K for 1 h.

Thus, the bands at 560 and 690 cm^{-1} observed in figures 3 and 2b are attributable to a surface Pt oxide species. Graham et. al. (16) have reported Raman spectra of all known oxides of Pt, although none of these resemble those of figure 2b. Therefore, it is proposed that the bands at 560 and 690 cm^{-1} pertain to a [PtCeO] surface interactive species in which Pt is bonded through oxygen to cerium. This is supported by evidence from TPR profiles of these catalysts (1,4,18). During reduction in hydrogen the removal of surface oxygen anions from ceria occurs at a much lower temperature in the presence of Pt. Both PtO and PtO_2 are reduced in hydrogen at between 423 K and 523 K. Thus it is possible that the lowering of the reduction temperature of surface oxygen by Pt is caused by a weakening of the Ce-O bond since this oxygen is also bonded to Pt.

If such a Pt-CeO_2 interaction exists then it might be expected that Pt would have a different oxidation state on a calcined Pt/alumina catalyst than on the catalyst containing ceria. Figures 6a and b show the IR spectra following CO adsorption on a freshly calcined Pt/alumina catalyst with a ceria loading of 20 wt. % and a Pt/alumina catalyst respectively. In figure 6a the predominant band at 2130 cm^{-1} is assigned to CO linearly bonded to Pt^{2+} sites, and the much weaker band at 2180 cm^{-1} is due to CO bonded to Pt^{4+} sites (17). In figure 6b the band at 2125 cm^{-1}, due to CO bonded to Pt^{2+} sites, is much weaker than the corresponding band in figure 6a and is accompanied by a shoulder at 2090 cm^{-1}, attributable to CO bonded to Pt^0 sites (17). The weaker band at 2180 cm^{-1} is also observed in this spectrum.

The observation of a band due to CO adsorbed on Pt^0 suggests the presence of particulate Pt on the Pt/alumina surface. Since no such band is observed for the ceria-containing catalyst and the predominant oxidation state thereof is Pt^{2+} it is likely that the stronger interaction between the Pt and the ceria, as compared to the interaction with the alumina, results in an oxidized, dispersed phase Pt site (19). This conclusion is strengthened by the work of Shyu and Otto (18). They observed, from XPS measurements that heating of Pt on alumina at 1073 K in air resulted in a surface species with oxidation state of Pt^{4+}, attributed to PtO_2. The same treatment of Pt on ceria yielded a Pt-ceria interactive species with an oxidation state of Pt^{2+}, attributed to PtO.

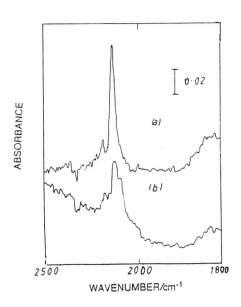

Figure 6. IR spectrum of CO adsorbed at ambient temperature on freshly calcined (a) Pt/alumina with a 20 wt. % ceria loading and, (b) Pt/alumina.

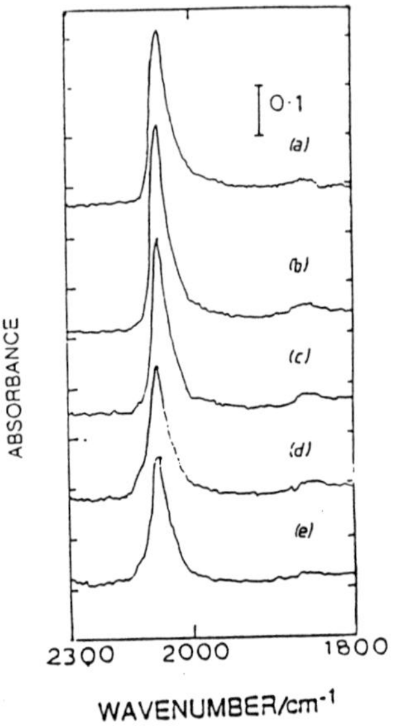

Figure 7. IR spectra of CO adsorbed at ambient temperature on reduced Pt/alumina having a ceria loading of (a) 20 wt. %, (b) 10 wt. %, (c) 5 wt. %, (d) 3 wt. %, and (e) 0 wt. %.

Figure 7 shows the IR spectra following CO adsorption on reduced catalysts having ceria loadings 20,10,5,3,0 wt. %, (a)-(e) respectively. A predominant band due to CO bonded to Pt^0 is observed in each spectrum. The CO band position on the reduced surface is independent of ceria loading, and therefore it is concluded that after reduction in hydrogen at 575 K the [PtCeO] complex decomposes.

CONCLUSIONS

(1) During calcination the $[Pt(NH_3)_4]^{2+}$ species decomposes and the Pt interacts with the acidic hydroxylated ceria to form a [PtCeO] complex.

(2) In this complex the oxidation state of the Pt is predominantly Pt^{2+}.

(3) During reduction at 575 K in hydrogen, oxygen is removed from this complex resulting in the breakdown of the [PtCeO] interaction.

(4) The formation of the complex is initiated when the alumina/ceria sol is impregnated with $[Pt(NH_3)_4]^{2+}$ at ambient temperature and is independent of ceria loading in the range 5 to 20 wt. %.

REFERENCES

1. H.C. Yao and Y.F. Yu Yao, J.Catal., 1984, 86, 254.
2. P. Loof, B Kasemo, and K.E. Keck, J.Catal., 1989, 118, 339.
3. T. Miki, T. Ogawa, M. Heneda, N. Kakuta and A. Ueno, J.Phys.Chem., 1990, 94, 339.
4. B. Harrison, A.F. Diwell, and C. Hallett, Plat.Met.Rev., 1988, 32, 73.
5. M. Ozawa and M. Kimura, J. Mat. Sci. Letts., 1990, 9, 291.
6. S.H. Oh and C.C Eickel, J.Catal., 1984, 87, 152.
7. Y. F. Yu Yao, J.Catal., 1984, 87, 152.
8. J.G. Nunan, H.J. Robota, M.J. Cohn, and S.A. Bradley, J.Catal, 1992, 133, 309.
9. Y.F. Yu Yao, Ind.Eng.Chem.Prod.Res.Dev., 1980, 19, 293.
10. B. Engler, E. Koberstein, and P. Schubert, Appl.Catal., 1989, 48, 71.
11. K. Nakamoto, " Infrared and Raman Spectra of Organic and Co-ordination Compounds", John Wiley and Sons, New York, 3rd Edition, 1978, Part 3, p. 199.
12. K. Otto, Appl. Surf. Sci., 1989, 37, 250.
13. J. Z. Shyu, W.H. Weber and H.S. Ghandhi, J.Phys.Chem., 1988, 92, 4964.
14. I.E. Wachs, F.D. Hardcastle, S.S. Unan, Spectroscopy, 1986, 1(8), 30.
15. S.S. Chan, I.E. Wachs, L.L. Murrell, and N.C. Disdpenzierie, J.Catal., 1985, 42, 1.
16. G.W. Graham, W.H. Weber, J.R. McBride, and C.R. PETERS, J.Raman Spec., 1991, 22, 1.
17. M.G.V. Mordente, and C.H. Rochester, J.Chem.Soc. Faraday Trans. 1, 1989, 85(10), 3495.
18. J.Z. Shyu and K. Otto, J.Catal, 1989, 115, 16.
19. H.C. Yao, M. Sieg, and H.K. Plummer, J.Catal, 1979, 59, 365.

Subject Index

Acetylene	191
Ag(100)	102,104
Al_2O_3	15,243
Aluminium Phosphate	202
Atom-probe field ion microscope	228
But-1-ene	101
But-2-enes	209
cis-but-2-ene	101
1,3-butadiene	145
Butane	98,276
Butenes	145
Butylidyne	101
C_2H_5CHO	130
CO	9,14,27,51,87
	99,101,128,146
	196,242,269,282
CO hydrogenation	234
CO oxidation	136
CO_2	14,46,89,167,184,222
Carbenium ions	4
Carbonate	24
Ceria	76,261,282
Chlorine	103
Cinchonidine	174
Cobalt	202
Copper	42
Cracking reactions	3
Cu(100)	101,213
Cu/Al_2O_3	16
Cu/Pd	165
Cu/Zn/Al hydroxycarbonate	63
Cu/ZnO	42
$Cu/Zn/Al_2O_3$	14,155
CuO	44
Cyclopropane adsorption	87
DRIFTS	127
EDX	249
EELS	97,238,242
EUROPT-1	176
EXAFS	46,51,203,255
Electron microscopy	249
Enantioselective hydrogenation	174
Ethane	129,191
Ethanol oxidation	221
Ethene	87,101,109
Ethene hydrogenation	207
Ethylene	128,191,207
Ethylene oligomers	3
Ethylidyne	101,109,112

$Eu_2Ir_2O_7$	184
Eu_2O_3	185
Exo-electron emission	190
FIM	229
Fischer-Tropsch synthesis	234
FTIR	87,98,118,145,165,196
Formaldehyde	25,163
Formate	22,119,170
Formic Acid	22,118,165
Gas Chromatography	127
Gd	79
Gd_2O_3	187
$Gd_2Ru_2O_7$	186
H_2	196
H_2 chemisorption	276
H_2O	123
HRTEM	196
HZSM-5	3
Hydride	200
Hydroformylation	127
Hydrotalcite	32
IR	97,282
Inelastic neutron scattering	68
Kinetic discontinuity	207
La	79
Lactate	174
Lanthanum oxides	76
Light off temperature (LOT)	277
Manganese	234
Mass spectroscopy	184
Methane	184,191,235
Methanol	155,203,213
Methanol conversion	4
Methanol synthesis	14,155
Methoxy	20,162,214
Methyl pyruvate	175
Microcalorimetry	87
Microwave	60
Mn/Ru(0001)	242
$Mo=CH_2$ complexes	87
MoO_3	256
MoO_3/SiO_2	87
Molecular beam	213
Molecular beam reaction spectroscopy	221
Molybdena-titania	255
Molybdena/alumina	68
N_2O adsorption	46
NEXAFS	213
NMR	1,34,155

Subject Index

$Na_2MoO_4 \cdot 2H_2O$	256
Nd	79
$Nd_2Ir_2O_7$	188
Nickel-aluminium mixed oxides	32
OH	123
Olefin metathesis catalysts	87
Oxidative dehydrogenation	213
Oxonium ion	11
Palladium	207,271
Pd(100)	99
$Pd-Fe/Al_2O_3$	145
Photoreduction	87
$(\eta-C_2H_4)Mo^{4+}$ complexes	87
Platinum	174,190,271
Pr	79
Propene	101
Propylidyne	101
Pt(111)	98,109
$Pt-Rh/Al_2O_3$	136
Pt/Sn	276
$Pt-Sn/Al_2O_3$	276
Pt/Al_2O_3	276
Pt/Rh	228
Pt/Rh/alumina/ceria	249
Pt/SiO_2	101
Pt/SnO_2	276
Pt/TiO_2	196
Pt/alumina	282
Pyrochlore-type catalysts	184
RAIRS	97,109,213
Raman spectroscopy	269,282
Rh(110)	221
$[Rh_{12}(CO)_{30}]^{2-}$	127
Rhodium	51,127
Ruthenium	234
Ruthenium-manganese bimetallic catalysts	234
SEM	191,249
SIMS	234
SSIMS	242
Silver films	118
Surface selection rules	98,121
Synthesis gas	184
T1 measurements	17,26
TEM	145,229,234,249
TGA	271
TPD	165,221
TPR	136,184
Temperature programmed decomposition	63
Thiophene	68
Tin oxide	270

Vinyl	112
Vinylidene	112
Water	155
XANES	203
XAS	42, 202
XPS	36, 76, 136, 234, 261
XRD	42, 64, 78, 184, 269, 276
ZnO	16, 44
ZnO/Al_2O_3	16

APR 16 '93